T0279789

# The Bullied Brain

## *Heal Your Scars and Restore Your Health*

JENNIFER FRASER, PhD

Prometheus Books

Essex, Connecticut

 Prometheus Books

An imprint of The Globe Pequot Publishing Group, Inc.
64 South Main Street
Essex, CT 06426
www.globepequot.com

Distributed by NATIONAL BOOK NETWORK

British Library Cataloguing in Publication Information Available

**Library of Congress Cataloging-in-Publication Data**

Names: Fraser, Jennifer, 1966– author.
Title: The bullied brain : heal your scars and restore your health /
    Jennifer Fraser, Ph.D.
Description: Lanham, MD : Prometheus, [2022] | Includes bibliographical
    references. | Summary: "In The Bullied Brain readers learn about the
    evidence doctors, psychiatrists, neuropsychologists, and neuroscientists
    have gathered that shows the harm done by bullying and abuse to your
    brain and how you can be empowered to protect yourself and others.
    It is not only critically important to discover how much your mental
    health is contingent on what has sculpted and shaped the world inside
    your head, but also the first step in learning ways to recover"—
    Provided by publisher.
Identifiers: LCCN 2021029929 | ISBN 9781633887787 (cloth ; alk. paper) |
    ISBN 9781633887794 (epub)
Subjects: LCSH: Bullying. | Bullying—Physiological aspects. |
    Bullying—Health aspects. | Victims of bullying—Mental health. | Brain.
Classification: LCC BF637.B85 F74 2022 | DDC 302.34/3—dc23
LC record available at https://lccn.loc.gov/2021029929

Printed in India

For my teachers, Montgomery and Angus Fraser-Brown

My brain isn't broken.

It's beautiful. I'm in a city I've never been to and I see bright
lights and new ideas and fear and opportunity and a thousand
million roads all lit up and flashing.

I say

There are so many places to explore but you've forgotten that
they exist because every day you walk the same way with
your hands in your pockets and your eyes on the floor.

—*Brainstorm*, a play by Company Three

# Contents

Foreword                                                                                   vii

Acknowledgments                                                                            ix

Introduction: Neurological Scars from the Bullying Paradigm                                xi

 1   Open Your Mind to the Twenty-First-Century Scientific Revolution      1
     *Step 1: Harness Your Neuroplasticity*

 2   Refuse the Lie That Abuse Is a Necessary Evil                            17
     *Step 2: Become a Writer of Culture*

 3   Learn to Disobey                                                          35
     *Step 3: Grow Your Talent*

 4   Exit the Cage of Learned Helplessness                                     55
     *Step 4: Unlearn and Rewire*

 5   Prevent the Cancerous Confusion of Bullycide                              71
     *Step 5: Grieve*

 6   Return Your Brain to Its Golden Potential                                 87
     *Step 6: Train Your Brain*

 7   Stop Identifying with the Abuser                                         105
     *Step 7: Believe in Yourself*

 8   Heal Your Brain with Your Mind                                           123
     *Step 8: Redraw Your Brain Maps*

 9   Restore Your Brain with Your Body                                        141
     *Step 9: Oxygenate Your Brain*

10   Listen to Your Mind-Brain-Body                                           159
     *Step 10: Hear Your Whole Voice*

Conclusion: The New *Neuroparadigm*                                179

Notes                                                              185

References                                                         207

Index                                                             221

# Foreword

Jennifer Fraser has written this book for you. You've been bullied or maybe you are or have been a bully; perhaps you care about someone who has been a bully or bullied. Alas, we humans abase one another on a routine basis. What most of us have not appreciated until recently is that these behaviors have serious, negative neurological consequences for both the bullied and the bully.

Dr. Fraser helps us understand the myriad of ways that we humans can deploy to humiliate one another. While child-on-child bullying is commonplace, so, too, is adult-to-child and (of course) adult-to-adult abuse. In all three cases, the bully or abuser proclaims their dominance over their "victims," feeding their egocentric appetite while they degrade their empathetic core. Most bullies are blind to the fact that their bullying and abusing behaviors are a strong form of progressive, self-inflicted neurological wounding. The plastic negative changes in that bully's brain grow apace and will be embedded in their personhood for life—unless something is done about it. Our modern culture is plagued by the endemic egocentric detachment that stems from all of this substantially avoidable social mayhem.

Dr. Fraser also helps the reader understand the even more direct neurological trauma that arises and progressively grows in the brains of the bullied or abused. In the past decades, the disabling impacts of fear and ongoing stress associated with bullying and other forms of child abuse have been extensively scientifically documented. Bullying induces "soft" brain damage. It impairs the progression of neurological advance. It carries the brain "off-line" in ways that negatively impact learning rates and personal achievement. It has a long series of negative consequences on organic brain health and on general physical health. It increases the risks of a progression to an anxiety or depressive disorder and of suicide. It increases the risks of a progression to addictions. These impacts are amplified, again, by that additional self-inflicted wounding that arises because the bullied and abused individual reevaluates and then enduringly stigmatizes himself or herself as a "victim," "weakling," "outcast" . . . a "loser."

Bullying has had consequences in Jennifer Fraser's life. She awakened to its happening in the lives of individuals that she cared deeply about. She witnessed its insidious neurological and physical impacts in people she loved. She explains how she has tried to tilt this enormous windmill. As a warrior, she did a wonderful job of preparing herself for battle by educating herself about the neurology and psychology and the cultural paradoxes related to this darker side of our human natures. As you read, you'll likely search your own mind and reexamine your own behaviors to reconsider how you personally govern your interactions with children and calibrate healthy child-to-child and adult-to-adult interactions. I'm a brain expert who is widely informed about the destructive neurological impacts of epochs of fear and ongoing stress in children and adults—and that is *exactly* what I did.

Most accounts of neurological and behavioral distortion arising from adverse childhood experiences end with an explanation of the environmental events that engender them, but Dr. Fraser takes a wonderfully insightful further step forward. In her own life and kinship, she was driven to answer a crucial further question: "Given the fact of bullying, what the h___ could we do to help the 'victim' move back on a path to neurological and physical normalcy?" Because she is herself an empathetic individual, it did not take her long to ask a second question: "What could and should be done to help that bully?" Again, our warrior entered this second arena after extensively educating herself about the science of brain plasticity-based neurorehabilitation. The brain of that current or historic bully or "victim" *is* (of course) plastic. Inflicted damage *can* be overcome—but healing requires that we acknowledge the wounding and take specific forms of action to drive the brain back in a corrective direction. Jennifer's and my dream is that our understanding of the neurology of bullying should and shall be succeeded by an epoch of personal and national healing.

Have you been a bully? Have you been bullied? Do you care about someone who has inflicted this form of neurological damage on someone and who, by such egregious behaviors, slowly but surely damaged him- or herself? Read this book to help yourself and to help *them* find a path to that better place where human empathy and positive inspiration and positive self-appraisal once again rule the day.

Michael Merzenich, PhD
Professor Emeritus, UCSF
Chief Scientific Officer, Posit Science
Founder and President, Brain Plasticity Institute

# Acknowledgments

I am privileged to work with literary agent John Willig, whose cordial expertise and unwavering faith in this project have been instrumental in bringing *The Bullied Brain: Heal Your Scars and Restore Your Health* to readers. Jonathan Kurtz, executive editor at Prometheus Books, and his team have been wonderful advisors each step of the publishing journey. Joe McNeely, senior editor at Brilliance Publishing, is to be thanked for transforming the book into an auditory experience.

I was privileged to discover passionate supporters of *The Bullied Brain* in Dr. Michael Merzenich and Wendy Haigh. They share with me an urgent and intense desire to see children released from the entrapment of the bullying and abuse framework. Dr. Merzenich lent his neuroscientific knowledge to the project, enhancing its accuracy and its breadth. That said, any and all errors belong to me.

As I was about to send the completed manuscript to Prometheus Books, I told my husband that he really should read it. He replied, "I don't need to read it. I live it." How true. We have gone through hell together raising a child with chronic illness and severe, persistent pain. Then with our healthy child, we found ourselves having to battle against a broken system to save him from systemic abuse and revictimization. Our sons have both been handed a challenging set of cards, but they have also been blessed by having a father who would take a bullet for them, and they've grown up grounded by this security.

I dedicated *The Bullied Brain* to Montgomery and Angus-James, as they have been my teachers. In their battle to survive they have drawn on such qualities as integrity faced with corruption; steely comportment faced with a smear campaign; mindfulness faced with agonizing, relentless pain; and wit, humor, and intelligence faced with the glaring limitations of the outdated, broken system that was designed to protect but that, in fact, enables abuse. Confronted with bullying and abuse, they found the courage to speak up when so many adults in their world looked the other way. My sons are not the exception; they are the rule. I have worked with young adults for twenty years in the arenas of literature and theater, and have learned a great deal from them. What stands out the most is their creativity and empathy, qualities that are missing in the bullying paradigm. Along with the brilliant insights of neuroscientists, children and youth are what give me hope and inspiration. Generations past have failed in a tragic way to halt abuse and bring healing and compassion to our world. If we let scientific study of the brain guide us, we have a chance to exit the bullying paradigm and enter into a new neuroparadigm. If we let children be our teachers, we can return to a more natural state of empathy, creativity, and care.

# Introduction

NEUROLOGICAL SCARS FROM
THE BULLYING PARADIGM

In 2012, we put our sixteen-year-old son, Montgomery, onto a flight from Canada to Kenya, which struck me—as I calmly waved good-bye—as an outrageous risk, but if I've learned anything since that day, it is that the brain can normalize *almost* anything. Montgomery was traveling with teachers and students on a service trip to build the foundation for a school where Kenyan children could have a more formal education, but still, what kind of parent allows their teenager to go halfway around the world to a country and continent they'd never even been to? Montgomery had attended an international private school on the west coast of Canada since he was in fourth grade, and I had taught at the associated senior school since that time. While local kids could attend, there was a significant boarding program that yearly hosted more than two hundred wonderful students from throughout the world. That international community of young people must have made it seem quite normal for our son to travel from North America to Africa, but for me, it felt wrong on a gut level to willingly send my child to the other side of the planet.

We had taken Montgomery to the doctor the day before because he had returned home from a school basketball tournament with the inside of his mouth and tongue covered in ulcer-like sores. He was in so much pain he could barely eat or drink. The doctor was surprised and asked him if he was "under a lot of stress." At this time, we had only heard brief, distressing details from the tournament about how the teachers, who were coaching the boys' basketball team, were putting them down, humiliating them, yelling in their face, berating and threatening, and just overall creating a fearful atmosphere, favoring some and blaming others. Further details about scenes of public shaming, the holding in for more yelling in the face, the repeated grabbing of our son when he tried to get away, the relentless swearing and homophobic slurs had not yet been documented. Still, the tournament sounded like misery. The doctor gave our son a prescription of antibiotics but told us that she thought his mouth had reacted that way because of "cortisol." I had not heard of the term before, nor had I looked up how "aggressive voices" affected the brain.[1]

I now know that cortisol is a stress hormone that courses through the body and brain when a person's sympathetic nervous system is triggered. This response to stress was carefully crafted by evolution to save our lives. While useful back in a time when we were faced with predators that required us to fight, flee, or freeze, its intensive effects designed for a short-term blast of adrenaline and cortisol were no longer helpful in most modern-day scenarios. In a situation where teachers are daily bullying children, a situation the victims can't really escape, cortisol works more like a corrosive substance, hurting both body and brain.[2] According to the Centers for Disease Control and Prevention, 80 percent of medical costs are stress-related, and as neurobiologist John Medina puts it succinctly, "This translates into a lot of cortisol."[3] It was painful as a mother to read that cortisol was not only burning the inside of our son's mouth but also interfering with "the executive center" in his brain and disrupting his brain's "developing architecture."[4] Was this just a few surges of cortisol at the tournament, or had he been suffering for a long time and it was only manifesting beyond his brain now?

I felt frustrated by my own lack of knowledge as a parent and as an educator. I was proud of being a published author and an award-winning teacher, but what good were my knowledge and skills when I knew nothing about the brain? How was it possible to parent, teach, and guide young people when you ignored the very organ that does the learning?

We didn't yet know the details about the teachers' abuse, but we knew enough that it was serious. We didn't want to ask Montgomery questions as his mouth was killing him and he had to get ready for his trip. I prayed that the tournament had been particularly awful but that the two years leading up to it, both with the same teacher, hadn't been all that bad. A brief influx of extreme stress was something you could recover from, but the more I learned about chronic stress, the kind you get when day after day you attend abusive practices, games, and tournaments, the more I worried our son was far more hurt than ulcer-like sores in his mouth. I have now learned that regardless of the tendency to dismiss and deny abuse, neuroscientists have found that when stress becomes repeated or chronic, the immune response doesn't turn off properly. This malfunction in the brain can lead to damage, such as nerve cell death, and when individuals are vulnerable, these changes can last much longer following the stress.[5] Was it possible that our son's immune system was now malfunctioning? Was he suffering from "nerve cell death"? And how long exactly would these "changes" to his brain last? These questions haunted me.

I read that when cortisol shoots repeatedly into the brain, the hippocampus (the memory and learning center of the brain) soaks it up because it is full of cortisol receptors. That's why if individuals are stressed out, they struggle to concentrate, make good decisions, and learn and retain information.[6] With our well-developed evolutionary survival strategies, the brain needs to remember specific dangers to avoid them in the future. The focus on danger interferes with the brain remembering all kinds of other details that don't seem as important. The brain that is chronically stressed-out is a brain focused on survival, pumped full of cortisol. Medina's list of the damage done by chronic stress was deeply concerning. In those who were repeatedly stressed, cortisol eroded their ability to do math or process language. It hurt their short- and long-term memory. It interfered with the ability to adapt information, concentrate, and learn.[7] I thought to myself: everyone needs to know this; certainly every parent and teacher needs to know this. We need to know this in kindergarten, and here I was a mother with a teenager, stumbling on this material, because my child was in pain.

I began looking for remedies, but it didn't take long to realize that, before I sought methods for healing scars and restoring health, a necessary first step would be to understand just how harmful bullying and abuse are to brains. The harm done to our son was invisible. He didn't have any cuts or bruises. How could we begin to understand the invisible impact on his brain from bullying and abuse, with its resulting chronic stress, when we couldn't even see it, let alone measure it? It was time to learn from the experts.

Neuroscientists study the way in which pain is designed to alert you to harm being done to both brain and body. Pain circuits in your brain get activated by a broken bone, but they also activate if you're excluded from a social gathering or a meeting at work. As doctor and addiction specialist Gabor Maté explains:

> The very same brain centres that interpret and "feel" physical pain also become activated during the experience of emotional rejection: on the brain scans they "light up" in response to social ostracism just as they would when triggered by physically harmful stimuli. When people speak of feeling "hurt" or of having emotional "pain," they are not being abstract or poetic but scientifically quite accurate.[8]

Our son was in extreme emotional pain, but we did not have a shared vocabulary to discuss it or know how to help him heal. While he was away, I learned that chronic stress is devastating to both the brain *and* the body.[9] Neuroscientist Stan Rodski's emphasis on how both brain and body were badly harmed by chronic stress confirmed what Medina was reporting: "Under chronic stress, adrenaline creates scars in your blood vessels that can cause a heart attack or stroke, and cortisol damages the cells of the hippocampus, crippling your ability to learn and remember."[10] If indeed our son was being abused regularly, and we wouldn't know until he returned from his trip, he had been in a state of chronic stress for at least two years. His brain was being damaged—the hippocampus—and his body was being damaged—the blood vessels. And it was all invisible and no one even talked about it, let alone informed parents, teachers, and kids about the risks. How could you take seriously and truly protect yourself from something that you knew nothing about and that no one seemed to see as a crucial safety issue? Yet social workers, who are on the front lines with traumatized kids, are aware that this health crisis has become totally normalized.

> If 20 million people were infected by a virus that caused anxiety, impulsivity, aggression, sleep problems, depression, respiratory and heart problems, vulnerability to substance abuse, antisocial and criminal behavior, . . . and school failure, we would consider it an urgent public health crisis. Yet, in the United States alone, there are more than 20 million abused, neglected, and traumatized children vulnerable to these problems. Our society has yet to recognize this epidemic, let alone develop an immunization strategy.[11]

The epidemic of abuse, neglect, and trauma has led to an epidemic of hurt brains. Abuse creates a cycle whereby the abuse victim's damaged brain becomes *abusive* either to self or others.[12] In the past, we could excuse normalizing this cycle of brain damage because we could not see the harm done to the brain by abuse. However, now that we can *see* the neurological scars on the brain with technological innovation, it's time to act. It's time to heal and restore our health. Every single individual who heals the trauma to their brain has a lower chance of suffering life-threatening health consequences—and has a lower chance of infecting and harming others.

Even more startling than society's failure to halt the cycle of abuse, neglect, and trauma is the fact that there is a *cure* for the 20 million infected children that is not being widely implemented. In fact, it's barely being implemented at all. Can you imagine discovering the cure for diabetes or cancer and then not applying it to those who are ill? *The Bullied Brain* dismantles the barricades that prevent our understanding of the impact of bullying and abuse on the brain; then it examines the series of cures that are well established in scientific research.

If brilliant scientific minds figured out that bullying and abuse can cause brain damage and that brain scans can reveal it, I had every hope that they would also have a host of remedies for recovery. And I was right. After each chapter that exposes how so many of us have been brainwashed in the bullying paradigm, there is a corresponding section on the remedies neuroscientists have discovered that heal brains and restore health. While you wade through the difficult sections on bullying and abuse, never forget that your brain is remarkably adept at healing when you train it according to extensive, evidence-based practices, recorded in neuroscientific research. The turning point of *The Bullied Brain* occurs halfway through. Chapter 5 is the final, most distressing analysis of the bullying paradigm. The second half of the book focuses on healing scars and restoring health. Chapter 6 offers the most exciting and targeted strategy for healing through brain training. Chapter 7 takes one final hard look at why it's so incredibly difficult to walk away from the outdated bullying paradigm, even when we are offered an evidence-based cure. The final three chapters offer inspiring further strategies for healing and restoring health.

The goal of this book is to change your mind using evidence. Once you see the way in which we've been sold a false bill of goods when it comes to the bullying paradigm, you will hopefully be emboldened to walk away from it and be empowered to seek an alternative framework. The alternative to bullying and abusive behaviors is absolutely within reach. All it requires is a change in mindset, based on learning a new way of thinking grounded in neuroscientific findings. In the labs of scientists, they discovered how harmful all forms of bullying and abuse are to our brains, and they have also uncovered a whole host of healing, restorative practices that we can undertake to recover from and prevent further bullying and abuse.

I use the term "bullying paradigm" as a shorthand for all of the inadvertent, as well as the purposeful, abusive behaviors that have become normalized in our society. While "bullying" is usually reserved to discuss behavior in child populations, I am going to apply it mostly to adults. Just as we are unlikely to make significant changes in economic disparity or environmental degradation by studying children, likewise, we do not solve the problem of bullying by focusing research on the young. *The Bullied Brain*'s focus is almost exclusively on adults who use all forms of bullying behaviors from microaggressions to physical violation.

Bullying applies to the tendency of parents, educators, or doctors to categorize and label children rather than see them empathically and treat them as holistic, complex beings with histories, hopes, and untapped potential. This normalized behavior is not meant to be cruel or destructive, but it can do significant harm to the child, or even adult, receiving the label. Oftentimes these labels are attached to learners without knowledge of their developing brains. Oftentimes these labels are attached to patients without factoring in trauma that has made the brain and body act at cross-purposes. Even mental health practitioners do not automatically assess the brain as an organ that rules body and behavior.[13] These kinds of labels belong to what I call the "bullying paradigm," which has become outdated since it was replaced by extensive neuroscientific findings. As psychiatrist Daniel Amen states: "Human behavior is more complex than society's damning labels would have us believe. We are far too quick to attribute people's actions to a bad character, when the source of their actions may not be their choice at all, but a problem with brain physiology."[14]

On the other end of the spectrum, I use the term "bullying paradigm" to encompass all forms of purposeful abuse including microaggressions and relational aggression. Microaggressions are so minor and subtle you might miss them. A microaggression happens when a sexist, racist, or homophobic joke is told; when someone speaks up and says, "That's unkind or inappropriate," and they're told to "lighten up. It was just a joke." A microaggression is when you tell someone about yourself or about an idea you have, and they don't respond. It's as if you never spoke. A microaggression occurs when you ask for feedback and it is not given. A microaggression occurs when you've been conversing with someone online and they don't reply, thereby ghosting you. These purposeful acts seem minor, but they do harm to your brain and they are the key underpinning of the bullying paradigm.[15]

Relational aggression involves bullying behaviors that attack relationships. Everyone is invited to a party or an important meeting except you. You arrive to school or work one day and people seem awkward and look away. You find out there's been a smear campaign meant to harm your reputation. Private exchanges you've had with someone, perhaps intimate images, are made public without your consent. Bullying that hurts your relationships is very damaging. While your body is not harmed, one of the most important aspects of a healthy and happy life, namely your social engagement, is harmed. Relational aggression does significant harm to your brain as documented in countless studies.

The term "bullying paradigm" encompasses all forms of abuse and sees many of them as intertwined: Emotional, verbal, psychological, physical, and sexual abuse don't occur in silos, but they are often treated as if they do. These behaviors can occur in person or online. They can attack the body and brain or just the brain. I also include in this set of abuses all forms of neglect ranging from physical to emotional. The umbrella term "bullying paradigm" is meant to establish that our culture and society have become so steeped in abusive behaviors that many of them go unnoticed. If they are noted, they often are denied or dismissed. If they actually can't be dismissed, the system lurches 180 degrees, and the one who reported or who was victimized all too often gets blamed, shamed, and ostracized (which are all bullying behaviors). We wring our hands about how our children are suffering from a bullying epidemic, but it's hardly surprising when these same children grow up in an adult world rife with normalized bullying.

Imagine how confusing it is for children. We tell them not to bully one another. We tell them they need to be "upstanders" and report to a teacher if they see anyone being bullied. We tell them that bullying emerges from a power imbalance that can happen among children even. Perhaps the victim is new at school, from a lower income, has talent that threatens the bullying child, is a visible minority, has some sort of challenge, or has a home life the bullying child envies. We are very clear about exactly what constitutes bullying, and we are vocal about how we have zero tolerance for it. The only problem is that all of these directives fail to admit, let alone address, that children *learn* bullying from adults. A more forceful way to express this and eliminate responsibility from child populations is: kids are trained by the adults in power over them—whether by role-modeling or more direct practices—how to bully. Children are not taught what constitutes *adult* bullying and abuse or how to handle the enormous power imbalance. If a child actually reports bullying, which is rare, they are often dismissed; the abuse is denied; they are faulted; they are told they deserved it; the abuse is renamed all kinds of different things like "discipline," "motivation," "passion," even "love." These adult responses to child abuse reports are extensively documented, but we continue to repeat the same platitudes while rampant abuse harms many children.

Headlines from the last few years suggest just how rampant abuse is even though we still don't protect children adequately or educate them how to stay safe from adults. If anything, adult abuse is treated like a taboo subject and children are left to fend for themselves. The following are just a few examples of such headlines:

- "Sex Offences against Minors: Investigation Reveals More Than 200 Canadian Coaches Convicted in the Last Twenty Years," with the subheading, "Expert says CBC investigation, 'tip of the iceberg,' calls for massive reform across the sport system"[16]
- "Why Does Women's Basketball Have So Many Coaching Abuse Problems?"[17]
- "Teachers who Sexually Abuse Students Still Find Jobs," with the subheading, "A year-long USA TODAY Network investigation found that education officials put children in harm's way by covering up evidence of abuse, keeping allegations secret and making it easy for abusive teachers to find jobs elsewhere"[18]
- "USA Gymnastics Culture of Abuse Runs Far Deeper Than Larry Nassar," with the subheading, "As the Tokyo Olympics loom, a pair of documentaries have exposed a decades long history of secrecy and exploitation"[19]
- "College of Charleston Hammers 'Jekyll and Hyde' Verbal Abuse by coach Doug Wojcik"[20]
- "Public Can Check Out Boy Scout 'Perversion Files' for Accused Molesters," with the subheading, "A Jacksonville man is joining a growing list of people suing Boy Scouts of America"[21]
- "The Inside Story of a Toxic Culture at Maryland Football"[22]
- "'It's Overwhelming': Survivors Create Public List of Catholic Clerics Accused of Sexual Abuse"[23]
- "Canada: 751 Unmarked Graves Found at Residential School"[24]

The concerning truth is that our system knowingly enables abuse, covers up abuse, and fails to stop it.

Adult abuse is rarely, if ever, discussed in conjunction with child bullying, which is odd because they form a well-documented cycle. In a survey of more than one thousand American educators, while it was found that only a few teachers and coaches use their position of trust and power to humiliate, harm, threaten, or induce fear and emotional distress in students, the effect is contagious and influences children's behavior, creating a widespread "harmful, discriminatory and hostile climate in which learning is undermined and intolerance flourishes."[25] Adults role-model bullying and children imitate them; it's just that we're not supposed to talk about it.

While for most, discussions around educators who bully is a taboo topic, sociologist Alan McEvoy shares that *his* interest in teachers who bully stems from his own "childhood experiences with a few teachers and coaches who waged a daily reign of terror over students." When presenting at schools on bullying, McEvoy repeatedly dares to discuss bullying by adults and has learned that the community of teachers and even administrators repeatedly express "a sense of powerlessness" when faced with a teacher or coach's "cruel behavior" to children.[26] It only takes "a few" teachers or coaches to create "a reign of terror" for children in classrooms or at practices.

In the workplace, adults also suffer from "workplace terrorism" when bullying goes unchecked.[27] The powerlessness appears grounded in fear. As lawyer and bullying expert Paul Pelletier advises: "These fears are real—you will be afraid of retribution, of making the problem worse, of not knowing how to speak up. However, it is silence and fostering the fear of speaking out that enables bullies to thrive."[28] Few want to tackle this troubling issue in public as it could oust you from the in-group. *The Bullied Brain* aims to shift this dynamic so that we use a new way of thinking and a new vocabulary to take our discussion of adult bullying and abuse out from behind closed doors and into a public arena where we can properly examine it, question it, and make changes in our own lives. The new way of thinking and new vocabulary are taken from the labs of neuroscientists and applied to our desire to heal our scars and restore our health.

Social work and management researcher Brené Brown pushes us to find the courage to enter into "the arena," which she describes as "any moment or place where we have risked showing up and being seen."[29] Those who take this risk have two qualities: "They recognize the power of emotion and they're not afraid to lean in to discomfort."[30] *The Bullied Brain* is an emotional book that may create feelings of discomfort, but it more than matches the weight of these feelings with the hopeful possibility of laying down a new path, a new collective neural network that leads us out of the bullying and abuse paradigm altogether and all together.

Without bringing neuroscience into the discussion, we appear incapable of change. McEvoy wrote about educator bullying six years ago. Little to nothing has changed since his forthright description of the cycle. In 2018, he wrote about how the learned helplessness of a victim morphs into adult powerlessness faced with "bureaucratic indifference" so that nothing ever changes in the bullying/abuse arena.[31] Although we regularly encourage children to *not* be "bystanders," both children and adults know that to speak up in a bullying or abuse situation runs the serious risk of being cast out. Whistleblowers quickly find themselves ostracized and punished for breaking with the in-group. Being in the out-group for adults, let alone children, is a dangerous place. As psychiatrist Helen Reiss states: "Empath indifference toward an out-group, taken to its furthest conclusion, can have life-or-death implications."[32] Anyone who has come up against the system, constructed by the bullying paradigm itself, knows that trying to protect victims and hold perpetrators accountable is a losing battle. I speak from personal experience.[33]

Who was I to try and take on the school administrators, police, lawyers, educational authorities, and government agencies, let alone the massive bullying paradigm itself? When you are defeated in battle, when your own son is brutally harmed and you failed to protect him, it's possible to feel hopeless and afraid. I was afraid of bringing even more harm to my son; I was afraid of failing; I was afraid I'd never get hired again, after resigning in protest; I was afraid that I didn't have the training or knowledge to truly translate the neuroscience into something that could bring about healthy, healing change. In a conversation about failure and loss, psychologist Peter Ciceri shared with me his thoughts on the archetypal hero's journey. He said "fear is the dragon" that the hero must fight to achieve his goals. Then he told me something that I have held onto tightly for the past ten years while I have engaged in what has felt like an utterly hopeless battle: "If you find the courage and skill to slay the dragon, then help will come."[34] There were many dark days, hopeless moments, errors, recollections of times I had failed my children or my students, times where I couldn't decode the science, days I couldn't find the words or the story to explain what I wanted. Then one day, the dark clouds cleared, the dragon fell from the sky, the sun rose, and there was suddenly help beyond my most extravagant hopes.

One of the world's leading neuroscientists, Dr. Michael Merzenich—professor emeritus at the University of California, San Francisco; chief scientific officer at Posit Science; founder and president of the Brain Plasticity Institute; and winner of the 2016 Kavli Prize, awarded for outstanding achievement in advancing our knowledge and understanding of the brain—offered to help me. It was a miracle.

When Merzenich shared with me his unpublished manuscript about his childhood, I understood why he was helping me. Looking back on his years growing up, he recognized that his brain was healthy. It had been "receiving high-quality information" and was ready to achieve at a high level at school. Yet, even as a six-year-old, the young Merzenich was aware of other kids not so lucky. He knew that far too many children "reach their 6 year old birthday with their brain in trouble." These children's brains have been underexercised or "their brains have been hijacked by stress or fear." Instead of becoming specialists at their passions—whether in sports, academics, or arts—these children become "specialists at protecting themselves emotionally." This hinders them at school and in life. Although not able to express this scientifically at an early age, Merzenich was aware that the "abnormality" in other children's brains stemmed "from the unfair burdens their brains have to bear" as they must "deal with their difficult early-life experiences."[35] While Merzenich had an idyllic childhood in rural Oregon, he's aware, especially after decades of studying the brain, that far too many children are "at risk for a future life of failure at school" due to their trauma. They won't just fail; they'll hate school. They'll develop an oppositional perspective and experience an unstable older life; they might be in trouble with the law, develop mental illness or addiction, and so on.[36] Even as a child—or perhaps *because* he was a child—Merzenich had an impulse to address this incredibly sad fate being dealt to children in his world. It has been the guiding star of his career and life's work. In his autobiographical story, he writes: "Mikey is going to grow up and be dedicated to doing something about this at an older age."[37] There's something powerful and fulfilling in hearing about a child who witnesses a problem in the world and sets their sights on fixing it and then as an adult becomes a pioneer in saving children's and adults' brains.

Just as I was suffering from a serious case of self-doubt, Merzenich read *The Bullied Brain* and told me that my "annotation focuses more on the behavioral than on the deep neuroscientific side (where there are tens of thousands of relevant citations)."[38] It is not necessary for nonscientists to read, let alone understand, the deep science. What we need to know is that it backs up the easier-to-understand and the easier-to-apply behavioral science. How reassuring to hear, from an international leader and expert, that enormous amounts of deep neuroscience support the message I wanted to convey. The science confirms what many of us know to be true and what has been documented by other disciplines for decades: all forms of bullying and abuse have the capacity and the likelihood to damage brains and shatter lives. More importantly, and less well known, the damage done to the brain can be healed.

I take full responsibility for all errors, flaws, and shortcomings in this book. That said, I am far more confident in what I am sharing with you since *The Bullied Brain: Heal Your Scars and Restore Your Health* has been rendered insightful and more accurate by a close reading and critical commentary provided throughout by Dr. Michael Merzenich. Henceforth, I will share with you his expertise, thoughts, and insights. I will refer to him as "Merzenich," which is simply to facilitate reading and does not convey the immense respect and gratitude I feel for him. For every blow and burn I suffered from the dragon, it was worth it to receive such remarkable help.

Before I met Merzenich, I had read his extensive research that documents the way "neurological noise," which he describes as a kind of destructive "chatter" frustrating the "brain's ability to record or recognize what's happening," can turn a healthy, functioning brain into one that is traumatized and can even suffer dementia.[39] I asked myself: is it possible that within the skull of a suicidal child, we would find a brain suffering from PTSD that is so degraded by "neurological noise" that it can no longer even support its own survival? Survival is the brain's most intense goal, and yet suicide is the second leading cause of death in youth populations. Isn't it time we started listening to the neuroscientists about what's happening to our brains?

Here is an example of neurological noise, the kind of chatter that has the capacity to confuse or rattle any healthy brain. Three teenagers decide that a child in kindergarten needs to be taught a lesson. Two of the teens hold the child down while the third teen hits the five-year-old with a blunt instrument until his legs are black and blue and he can't walk. The kindergartener needs to go to the hospital because the internal bleeding is so serious. Now, if you feel very sure that this is bullying, imagine the same scenario, but replace the teenagers with adults in positions of trust, power, and authority over a fourteen-year-old. Is it still bullying? I would argue that it is unquestionably bullying. In fact, it should be categorized as child abuse and criminal. It should, without a doubt, be identified as cruel and unusual conduct on the part of three teachers, but there are many authorities in positions of influence and power who disagree.

In 1977, three teachers in the United States punished a fourteen-year-old boy at school. Two of them held him down, and the third one beat his legs with a paddle until they were black and blue and he couldn't walk. His parents took him to the hospital and launched a lawsuit against the teachers. The judge ruled that the teachers' conduct was *not* "cruel and unusual" and noted that, while there was a law that prohibited those who worked in jails from beating an inmate in this way, it did not apply to teachers and students. This was a precedent-setting case, and it continues to influence nineteen states in America where it is legal for teachers and administrators to hit children with paddles.[40] This is the reason I refer to a "bullying paradigm" and am taking back the word *bullying* from the confusing and slippery way it is too often used by adults.

Approximately ten to twenty thousand students in America "need medical attention as a result of being disciplined with corporal punishment in school, each year."[41] A history of remarkably brutal whipping of children with birch, bamboo, and other objects at hand is extensively documented in independent schools in Britain.[42] Schools in Canada were banned from corporal punishment with the strap in 2004, but parents were enabled by the Supreme Court to strike their children legally.[43] For more than fifty years, psychologists and psychiatrists have provided ample evidence that corporal punishment not only hurts bodies; it hurts brains. This is now confirmed via the noninvasive technology of brain images, but the law and the outdated mindset it reflects are slow to change. The belief system is still firmly in place despite statistics revealing a surprisingly disordered and underperforming youth. "Corporal punishment is ineffective as a form of discipline. It is correlated with violence, mental health concerns, and other problems associated with juveniles, such as lower levels of achievement and physical injuries."[44] American youth, despite the wealth and resources of their nation, trail the world in terms of health and performance. Economist Jeremy Rifkin writes: "Inflicting corporal punishment on a child for a social transgression only serves to lessen a child's empathy for others."[45]

American independent schools are permitted in forty-eight states to use corporal punishment. The government does not track the impact on students, their health and well-being, or their academic and behavioral success. In fact, there is concern that private schools "may be using corporal punishment to a far greater degree than publicly known."[46] Abusing a child's body or brain does *not* get results, but according to extensive research, it does do unspeakable harm. You cannot help but wonder if there is a correlation between the failure to protect children from corporal punishment in some U.S. schools and how poorly American students are faring when compared to other countries with far less resources. Psychologist Laurence Steinberg expresses frustration and concern:

> When a country's adolescents trail much of the world on measures of school achievement, but are among the world leaders in violence, unwanted pregnancy, STDs, abortion, binge drinking, marijuana use, obesity, and unhappiness, it is time to admit that something is wrong with the way that country is raising its young people. That country is the United States.[47]

These indices of trauma don't stop at high school; they spill into postsecondary education as well. Steinberg refers to statistics that reveal bullying, self-harm, eating disorders, drug use, attempted suicide, and the need for remedial education among college freshmen are on the rise. Studies in the last decade show that spanking, let alone hitting a child with a blunt instrument or strap, lowers a child's IQ and erodes developmental growth. Recent research done in eastern Canada found that spanking reduced the brain's gray matter, the connective tissue found between brain cells. Extensive research reveals that children subjected to adult bullying have less gray matter in the brain than those who have not been mistreated. Gray matter is integral to muscular control, sensory perception, speech, emotions, and memory, all arguably crucial to success.[48]

The term "bullying" applies to *all* cruel and unusual behavior, regardless of adults who apply the term selectively to children and switch it up when it's the adults who are cruel and unusual. Teachers may argue that when they strike a child with a blunt instrument, while having a significant power imbalance over the child, it is not bullying; it's "discipline." When a parent, teacher, or coach lambastes children in demeaning ways, yelling, shaming, threatening, they'll tell you it's not bullying; it's "passion." They got carried away because they care "too much" for the children. When adults groom children and lure them into sexual relations, they'll tell you it's not abuse; they "love" the child. These Orwellian reversals are refused in this book and are replaced by the term "bullying" to refer to all the forms of abuse that are passed on through generations to children who are told not to do them. Notably, Orwell was himself a victim of a highly abusive school system.[49]

Although many believe that physical harm is worse than emotional, research documents that executive function in the brain—making good decisions, considering consequences, self-regulation, ability to concentrate, impulse control, and so on—is jeopardized equally by both physical *and* emotional abuse.[50] A reduction of gray matter was seen in prefrontal areas of the brain where executive function occurs, regardless of whether a teenager had been physically or emotionally abused.[51] Gray matter influences intelligence and learning as measured on tests. Extensive research done by Matthew Lieberman and Naomi Eisenberger, which has now been replicated, reveals the harm physical and emotional bullying does to brains. Their work stresses how physical and emotional pain both leave lasting scars in the brain.

> Studies from around the world, including the United States, England, Germany, Finland, Japan, South Korea, and Chile, suggest that between the ages of twelve and sixteen about 10 percent of students are bullied on a regular basis. Although bullying can involve physical aggression, more than 85 percent of bullying events do not. Instead, they involve belittling comments and making the victims the subjects of rumors. But victims of bullying suffer long after school is over and the bully has gone home.[52]

How does their suffering manifest beyond the brain where we can't see the scars? Depression, suicidal ideation, and suicides.

In the bullying paradigm, adults use corporal punishment, along with blaming, shaming, and ostracizing to change children's behavior. It is backward thinking because all of these behaviors harm brains, and it is brains that manage behavior. Thus, much as we may wish to punish, blame, and shame perpetrators, we have to recognize that in many cases, perpetrators, who struggle to manage their behavior, were once victims. A glance at California's prisons exposes the cycle as 70 percent of prisoners spent time in foster care while growing up.[53] Prisons are full of bullied brains. Psychology and psychiatry have compiled decades of documentation that "many of the emotions, beliefs, and reactions" that trauma victims experience are in fact "unmetabolized responses to trauma of the past." These responses go on to negatively impact relationships and learning.[54] Perpetrators, who do harmful things, more often than not have bullied brains due to past trauma from abuse, as documented by brain scans.[55]

Even when perpetrators are children, we treat their poor conduct not as a brain issue but as "bad behavior" resulting in corporal punishment, suspension, or expulsion. Extensive research is clear that the "school-to-prison" pipeline simply piles trauma upon trauma, making it more and more difficult for brains to heal.[56] We are more likely to halt the cycle or the contagion of bullying by striving to cure and heal hurt brains, rather than punishing, blaming, shaming, and ostracizing them. Psychiatrist Daniel Amen writes: "Our work taught us that **many people who do bad things often have troubled brains . . . that was not a surprise . . . but what did surprise us was that many of these people had brains that could be rehabilitated**."[57] Amen bolds statements in his books when he really wants to grab his reader's attention, and the ellipsis points, I believe, are meant to convey a process, the process of discovery in his work on brains. Rehabilitating hurt brains does not mean that victims need to be exposed. Protecting victims must be paramount, but healing brain disorders, or rewiring the neural networks that lead to destructive behaviors, can occur in far healthier ways.[58] Perpetrators are more likely to self-identify and to seek help as early as possible if they are taught that their problems are medical, not moral, and that they can be healed and returned to health. "**Imaging immediately decreases the stigma as people begin to see their problems as medical not moral. We have nothing else in psychiatry that is this powerful or immediate**."[59] The bolding should tell you how strongly Amen feels about this.

We have entered into a whole new era due to neuroscientific findings. They are the antidote to the bullying epidemic. They have provided us with a way to dismantle the bullying paradigm and replace that broken, outdated framework with a new one grounded in knowledge of our brains. The overarching goal of *The Bullied Brain: Heal Your Scars and Restore Your Health* is to share neuroscientific findings on how bullying and abuse impact the brain, in ways that are enlightening but that also pave the way to practical application in our lives. Knowing your brain has suffered from bullying and abuse is a crucial first step in healing your scars and restoring your health.

We must first spend some time examining how we are brainwashed into believing that bullying naturally surfaces among children and is a part of growing up. We need to stop believing that adult-to-child bullying is a necessary evil to attain greatness. It is time for us to recognize that the systems put in place to stop bullying and abuse are broken. Once we debunk the myths that prop up the bullying paradigm, that we have been raised in and struggle to even see or question, then we can look more clearly at the evidence-based ways in which the bullied brain can become stronger, healthier, and happier. While it is challenging to read about the devastating impact of bullying and abuse on our brains, it is also incredibly exciting to learn just how skilled our brains are at healing and recovery. As Merzenich's decades of research have shown, you have a "rather astounding human capacity to change" because "you have the power of transformation" wired right into your brain.[60]

At sixteen years old, his mouth raw with corrosive cortisol, Montgomery packed the antibiotics into his suitcase. Within twenty-four hours, he contacted us from Kenya to tell us that the sores in his mouth had healed and he didn't need to take the medication. I was struck by how much the body reacted negatively to a repeatedly abusive environment, and yet how quickly it healed when removed from such toxicity. It made an impression on me. I was only just learning about the harm to my son's brain, but the inseparable nature of body and brain meant if the one could heal, maybe, just maybe so could the other. This crisis with my son led me on a journey where I learned that neuroplastic approaches to healing "require the active involvement of the whole patient in his or her own care: mind, brain, and body."[61] The key word is *whole*, the whole patient, not a broken self but a holistic self, and that meant the alignment of Mind-Brain-Body.

I learned that the way to exit the learned helplessness of the bullying paradigm is to bring your Mind-Brain-Body into awareness and even dialogue, supporting and caring for one another, rather than being fragmented and working at cross-purposes. As neuroscientist Norman Doidge explains: "The word heal comes from the Old English *haelan* and means not simply 'to cure' but to make whole."[62] The bullying paradigm, which is so normalized we don't even see it anymore, even influences Western medicine. Doidge contrasts the neuroplastic approach to healing or making whole with the "divide and conquer" medical approach, whereby the patient's body is "less an ally than the battlefield, and the patient is rendered passive, a helpless bystander" in the battle waged by the doctor against the disease.[63] In the neuroplastic approach to healing in this book, you are encouraged each step of the way to resist and reject the entrenched beliefs of the bullying paradigm, to take active charge of your own holistic health, and to understand that you are anything but helpless.

When Montgomery returned from Kenya, he finally told us about the relentless abuse he had been suffering from the two teachers coaching basketball. More and more students came forward to provide detailed testimonies about abuse that had gone on for years. A talented and award-winning athlete at any sport he set his mind to, Montgomery had chosen basketball to pursue at college and he was skilled enough to make that dream a reality; however, he told us that he had decided early in the season to set one goal and that was "not to let them break me." While Montgomery will never play basketball competitively again, he succeeded in remaining whole. This sadly is not the case for many victims, and that is why *The Bullied Brain: Heal Your Scars and Restore Your Health* is needed.

For a *Scientific American* article, journalist Rachel Nuwer chose the title "Coaching Can Make or *Break* an Olympic Athlete" (emphasis mine).[64] The title struck a chord with me. The key thing to remember is, if you're feeling "broken," never forget that you have the tools within to repair. Even a brain that has suffered massive brain damage can repair. As neuroscientist Sarah-Jayne Blakemore and psychologist Uta Frith teach, the brain has "resilience" that manifests in neurons that "can start the process of repair."[65] If you have been "coached" in your life by a parent, teacher, partner, manager, boss, or mentor who was bullying or abusive, then you might need to heal your scars and restore your health. If you weren't parented, taught, or coached to fulfill your amazing potential and instead you were "broken," your brain is primed to reverse the harm done. If your belief in yourself wasn't ignited but extinguished, your brain is designed to return that belief to you and set in motion a transformation. If your talent wasn't grown but withered, your brain has the innate capacity to regrow new, healthy neural networks, bringing your talent back online.

I did my master's and PhD in comparative literature at the University of Toronto. This form of academic training encouraged us to bring disparate ideas, languages, cultures, lenses, and discourses together to create new perspectives, understandings, and dialogues. How does our understanding of Shakespeare's plays change when viewed through an economic lens? How might psychological studies of trauma open up deeper levels in Hemingway's novels? How does knowledge about the brain change our understanding of teaching, coaching, and parenting?

My first book, *Rite of Passage in the Narratives of Dante and Joyce*, examined Italian and Irish literary giants, one from the medieval era and one from the modernist era. They couldn't be more different. I then placed their poetry and prose into an anthropological and sociological context. The goal was to explore rite of passage and to pose questions around why some individuals went through a difficult initiation to become writers of culture. The vast majority of us remain readers of culture: we absorb or learn and then think, feel, and work within the framework created by our culture. But there are those independent thinkers, those creative renegades, those rebellious, courageous, maybe even outrageous figures who walk away from the established way of thinking, feeling, and working, and they write into the world a whole new way of being. They are the textual equivalent of the Impressionist painters who changed the way in which we experience light or the Group of Seven painters who had us reimagine our relationship to nature. This is the literary or artistic version of what Thomas Kuhn defined as the work of "extraordinary scientists," the ones who exit the established paradigm and bring into a being a whole new scientific perspective.

My second book, *Be A Good Soldier: Children's Grief in English Modernist Novels*, was even more wide-ranging in that it drew from British novels in the modernist era to show the concerns writers had with contemporary education and parenting models. These pedagogical models demanded that children suppress their grief and "be good soldiers." It fascinated me that so many diverse authors, ranging from Joseph Conrad to Virginia Woolf, were writing fiction that made a connection between requiring children to dry their tears and the numbing (internal) or war-mongering (external) behaviors that dominated that time and led to the First World War and Second World War. I read a great deal of nineteenth-century pedagogy that advised parents and teachers to use harshness with children as they were seen as willful, unruly, and a threat to adult control. These same models were enacted abroad as England, seeing itself as the Great White Father, colonized other countries with notable cruelty. Juxtaposing cruel child-rearing practices with psychological study into how grief suppression leads to numbness or aggression provided a whole host of insights into the psychological and political concerns.

My third book, *Teaching Bullies*, told the story of what happened to my son Montgomery and many other students at the private school where I reported the teachers' abuse, setting their experiences in the context of extensive psychological, psychiatric, legal, neuroscientific, and abuse research. The story was also covered by an award-winning investigative journalist for the country's most widely read newspaper, as well as featured on a television program that uses investigative journalism to expose corruption and cover-ups.[66] While it was important for Montgomery to speak up publicly about the abuse and the school administrators' victim blaming, the larger issue in the media was the government's cover-up. By this point in my ongoing quest to figure out why so many universities, clubs, sports organizations, schools, and workplaces failed to stop abuse and instead enabled it and then covered it up, I knew that even this exceptional media coverage would make zero change. It's part of the pattern: abuse occurs and usually goes on for a number of years or even decades, cover-up ensues, abuse and cover-up are exposed in the media, and then, after a flurry of promises and firings and carrying on, we go back to business as usual until the next scandal is reported in the media. That was the pattern I wanted to understand and to break.

Danish author of fairy tales Hans Christian Andersen wrote a story in 1837 about a pair of swindling tailors who promise to create a beautiful, elaborate set of new clothes for a proud emperor. It is all a lie. The tailors cheat the emperor and create absolutely nothing, all the while convincing the emperor of the shimmering golden threads of his new outfit. The fictional clothes are completed, the fooled emperor believes he's dressed in them, and naked, he parades through his people. Not one adult dares speak the truth or question the emperor, for fear of igniting his wrath and being locked up by his guards. With his flatterers gushing about how exquisite his new clothes are, the emperor's subjects will themselves to see his new "clothes," and they clap and bow until a young boy speaks up and says, "The emperor wears no clothes."

In *The Bullied Brain: Heal Your Scars and Restore Your Health*, we speak about the golden threads of the bullying paradigm as a dangerous fiction, even if it makes us feel afraid. We use science to find the courage of a child and get to the naked truth.

# CHAPTER 1

# Open Your Mind to the Twenty-First-Century Scientific Revolution

## *Step 1: Harness Your Neuroplasticity*

When I was four years old and my brother was in first grade, he came home from school one day and said to my parents, "I think my brain is crippled." This resulted in him being whisked off to a child psychiatrist who diagnosed the issue as my parents not providing us with a full-length mirror that would allow us to see our whole selves. The psychiatrist diagnosed my brother as believing that his "self" ended at the waist. This was my first inkling that maybe the experts didn't always get the diagnosis right. More relevant for us as kids was that maybe teachers didn't always get it right. Maybe teachers didn't always pass accurate judgment on our potential. Maybe teachers lacked knowledge or the wherewithal to properly assess our ability to process information and learn skills. My brother's endless reading, his photographic memory, his rich vocabulary, his humor, his passion for history and literature did not gel with how he was treated or graded at school. His brilliance did not protect him from bullying by his peers. Watching the way my brother suffered in the education system, bullied inadvertently by teachers and with a vengeance by some children, gave me a keen sensitivity about learning exceptionalities at an early age. It might have been where I developed my feelings of resistance toward teachers who ruled students as if they were gods. Little did I know that this rebellious streak would save me in high school.

It wasn't until seventh grade, after years of suffering in the school system, that my brother encountered a teacher who didn't tell him he "wasn't trying" or "was careless and lazy." He was so knowledgeable when answering questions and debating in class but couldn't produce his excellent ideas in handwritten form. This was before laptops. This was an era when even typewriters weren't a part of school. In the 1970s, we were raised with the belief that handwriting and math were the most accurate indicators of our intellectual capacity and future potential. If you could write tidily and stay in the lines, you were seen as gifted.

I still remember the name of my brother's seventh-grade teacher, Mr. Bowman, because he recognized that my brother's brain processed information in a unique way and he treated him with respect and empathy. Mr. Bowman's belief in my brother's striking intelligence was enough for my brother to believe in himself and find ways around an unimaginative and oftentimes cruel school system. After seventh grade, my mom wrote to Mr. Bowman every year to tell him that my brother was popular and doing well academically in high school; that my brother was graduating from university; that my brother's off-off Broadway play in New York City was making waves; that my brother's miniseries for TV had been written up in glowing terms in the *New York Times*. What if there hadn't been a teacher like Mr. Bowman?

You would have thought that with the history I shared with my brother, I would have been extremely sensitive to and aware of any kind of unwitting harm being done by teachers to my own sons, let alone outright abuse. While my brother had teachers who didn't know about challenges like ADHD or dysgraphia and thus could be forgiven for their ignorance, Montgomery and his teammates were being called "f**king retards" by one teacher, while another one, who preferred homophobic slurs, would stand by and watch. Montgomery has dysgraphia (learning exceptionalities are often passed on in families), and the teachers yelling these kinds of insults were fully informed about his need to be on an IEP (individual education plan) that accommodated the way he learned. While Montgomery had been on the Headmaster's Honor Role for several years due to his intelligence and hard work, he was still sensitive to a term like "retard." Moreover, as an auditory learner, he was hypersensitive to yelling, especially when it was inches from his face.

As early as the 1990s, educators were writing about the way in which a stressful, emotionally unregulated, and threatening environment puts the hippocampus—a crucial center for learning—and indeed the whole brain on high alert. Learning and skill development come to an abrupt halt as the brain tries to cope with the threat: "Under perceived threat, we literally lose access to portions of our brain."[1] Trying to figure out the harm that might have been done to Montgomery, I Googled "brain and bullying," "emotional abuse and physical abuse," "neuroscience of bullying." Unlike most people, I had been introduced to neuroscience through the brain training in which Montgomery's younger brother was engaged. I had a layman's sense of "neuroplasticity" and was therefore curious what impact emotional abuse might have on the developing brain.

While trying to see what the research said about the impact of abuse, I came across one article that completely changed my understanding and raised a lot of disturbing questions: "Inside the Bullied Brain: The Alarming Neuroscience of

Taunting," written by Emily Anthes and published in the *Boston Globe* in 2010. My heart started to beat faster as I read through the article. Anthes noted that professor of psychology and education Tracy Vaillancourt's research findings showed that bullying can weaken the immune system and "kill neurons in the hippocampus."[2] I was unaccustomed to the term "neuron," let alone "hippocampus." I learned that a neuron is a brain cell, and the hippocampus is an area of the brain involved in memory storage and recall. This foreign language that allows us to better understand our brains is surely as important as learning French, Mandarin, or Spanish. Why are we not taught in school the meanings of the words that allow us to see into our skulls and become informed about our brains?

As a parent, I needed to know that the bullying from two teachers was raising my son's cortisol levels to dangerously high levels, just like I would need to know if these teachers repeatedly broke his arm. But while he didn't even have a vocabulary with which to articulate the invisible injuries to his brain, a broken arm would be instantly attended to and protected in a cast by medical experts so the bone could mend.[3] Writer James Clear uses "atomic habits" to lead a healthy life, a life threatened when a kid throwing a baseball bat accidentally hit his head. While Clear as a child was rushed to the hospital, flown in a helicopter to a more sophisticated hospital, and met by twenty medical personnel running to save his brain, the bullied brain does not get the same kind of medical or social recognition, let alone healing. Clear was harmed once by accident, whereas the bullied brain is harmed by purposeful actions over and over again.[4] It's important to make this distinction because those who were bullied tend to beat themselves up about how they are failures, how they can't stop addictive or destructive habits. Before you can apply Clear's brilliant "atomic habits," you need to learn about how your brain is injured; society won't do it for you. We have no problem intervening to heal the body; why do we not have the same urgent and effective approach when harm has been done to the brain? Anthes's article was upending whole constructs that I had lived by, and had not questioned, as a parent and as a teacher.

Vaillancourt had been frustrated in the past because her research into bullying and its serious impacts was largely ignored; however, now that she was able to offer biological evidence, her work was making headlines. Ten years ago she was "hopeful" that this new way of understanding harm done to the brain would be a "policy changer."[5] As far as I can tell, nothing has changed. Over ten years of working double time to see policy change and widespread recognition that we are operating in what Merzenich refers to as a "bankrupt notion," I have discovered what an uphill battle it is to make those in education, in health, and in power understand how destructive bullying behaviors are to brains, and yet how gifted our brains are at changing and healing if properly cared for and trained.[6] Even award-winning, at the highest level, world-renowned neuroscientists like

Merzenich have *not* been able to bring about this change. What he refers to as a "revolutionary scientific perspective" surely should now inform our lives in constructive and healthy ways, and yet it is largely ignored and unknown.[7]

Merzenich told me that at about the mid-century mark, "the medical-neuroscience mainstream began to conduct studies that seemed to show that brain remodeling was limited to an early post-natal epoch." Scientists thought that after early childhood, brain wiring was installed for life. However, physiological psychologists were studying behavioral conditioning, better known as "Pavlovian conditioning," that revealed the "capacity for older brains to change." These studies were mostly ignored by mainstream brain science. Merzenich's work takes off at this point: "My research team, operating in that mainstream, conducted compelling studies that showed, in substantially more complete and more useful ways, how the brain was remodeled all across our lifetime by our experiences." This groundbreaking research "very extensively documented both the positive and negative dimensions of those changes in the domains of place, time, and intensity." Merzenich and his team designed and conducted research "to provide the basis for 'controlling the Genie' in the form of practical applications to 'address human children and adults who struggle.'"[8] In chapter 3, we will go over this history in more detail as it's crucial to understanding our brains and their capacity for change.

When Merzenich realized that the adult brain had neuroplasticity, which meant there would be many opportunities for "strengthening and recovering brains," he assumed that the "scientific and medical communities would quickly correct their wrong-headedness." Instead, he was met with skepticism and subjected to "substantial ridicule." After traveling the long road required to re-model "the brains of all those brain scientists who were stuck" in an entrenched mindset, Merzenich explains that now most scientists understand that the brain is plastic from birth until our final days on the planet.[9] Scientists understand this critically important, revolutionary concept, but how many others do? The revolution happened, but it has barely extended beyond the confines of scientific labs and into the mainstream where it is desperately needed.

On that fateful day when I sat reading Anthes's article, I did not know that I would come to share Vaillancourt's frustration or Merzenich's battle to change entrenched beliefs. I had no idea that it would be a long journey to get others to recognize just how harmful the outdated bullying paradigm was and how much research had been done to expose it. Watching my son strive to heal his neurological scars made me realize that while we *can* heal from bullying and abuse, society's failure to support this healing and restoration of health makes it a much more burdensome and difficult recovery.

At the outset of Anthes's article, she writes: "Bullied kids are more likely to be depressed, anxious, and suicidal. They are more likely to carry weapons, get in fights, and use drugs."[10] When you're reading about this research and it applies directly to your own child, you hear the words differently. You slow down until you find yourself hanging on every word: depressed, anxious, and suicidal; weapons, fights, drugs. Montgomery has been described many times with the phrase "still waters run deep." I knew he would not act out in aggressive ways. He was far more likely to turn the repeated blows he had suffered into internalized misery. And indeed he became highly anxious. He suffered a major depression. He began to have panic attacks. These manifestations of what bullying does to brains did not emerge only from the daily abuse dished out by his teachers. These reactions also surfaced when the headmaster revictimized the students who spoke up. This betrayal devastated our son, until he made a conscious decision to put the whole thing in the rear-view mirror and get on with his life. I would like to inform you that this is a rare occurrence, that administrators in positions of power to protect children most often fulfill their mandate, but I can't. I have learned in my extensive studying of child abuse that it is *normal* for administrators to protect perpetrators and revictimize those who report.[11] If substance abuse, weapons, fighting, anxiety, depression, and suicidal ideation are the outcome of bullying and abuse, why don't we stop it? This question would drive me for the next ten years to find answers. The most enlightening and practical answers emerged from neuroscientific research.

The initial focus in Anthes's article was on peer bullying, from one student to another, but what about influential adults? What about parents, teachers, and coaches? All I could think was, if an essentially powerless kid could have this effect on a victim's brain, how much worse would it be if the bully or bullies were teachers who had immense power over a teenager? Bullying hinges on power imbalance, but the greatest power imbalance, namely adult to child, was rarely, if ever, even discussed. At first, Anthes talked about how we tend to dismiss bullying because it leaves no "obvious injuries" and essentially is understood as just hurting a victim's "feelings." But then she went on to say that a "new wave of research" was revealing that "bullying can leave an indelible imprint on a teen's brain at a time when it is still growing and developing."[12]

I was starting to develop a pit of fear in my stomach. What did she mean "indelible imprint" being left on the brain? A kind of imprint or stain that could never be washed away? I read on, at turns furious and worried. How dare these teachers do this kind of damage to teens who are in the throes of development? And yet even then, I had to admit to myself that we teachers had not been told about this "new wave of research." I myself knew next to nothing about the brain, let alone the developing brains of adolescents and how vulnerable they were.[13]

On the second page of Anthes's article, I found what I had been looking for but also dreading. Professor of psychiatry Martin Teicher conducted a study of one thousand young adults to see if verbal abuse by adults harmed brains. His research revealed that "verbal abuse could be as damaging to psychological functioning as the physical kind." He then wondered if peer-to-peer verbal abuse, namely "teasing, ridicule, criticism, screaming and swearing," had a similar impact on young adult brains. The research team conducted sixty-three brain scans of bullied youth and discovered they could see bullying had impacted their "corpus callosum," which had less insulation or myelin than healthy subjects.[14] Again, this was foreign vocabulary for me. I learned that the corpus callosum is made up of tightly woven fibers that connect the right and left hemisphere of the brain; it's involved in visual processing and memory, along with many other key functions. The more "myelin" or insulation it has on it, the more efficiently it operates. This is just the beginning of what neuroscientists have learned about the negative impact of bullying and abuse on brains. These are the kinds of findings that we will look at throughout *The Bullied Brain* because you, like me, will struggle to "heal your scars and restore your health" without having any idea about the harm that may have been done to your brain or the brains of others in your life.

Despite reading and critical analysis being what I most excelled at, in all honesty, I couldn't take in Emily Anthes's article all at once. I felt betrayed and frustrated. Why did professionals in the business of education know next to nothing about how to keep children's brains safe and healthy? I sat back in my chair and stared out the window. I was shocked by my own ignorance about something so important. In that painful moment, I decided that I would educate myself on the impacts of bullying and abuse on the brain. I would share my knowledge with my colleagues every step of the way. We would launch a movement that would draw the attention of educators worldwide on how vulnerable the brain was to bullying. Coaches and teachers have enormous influence and can be positive change-makers. During my vocal advocacy, I have found that internationally a movement is afoot to oust the bullying paradigm and bring in a more informed, nurturing, empathic, supportive, mindful framework. It cannot come soon enough.

Only in recent years have we learned from brain scans that concussions are serious and dangerous.[15] Concussions cause life-threatening injuries to the brain, and while it can heal, it must be treated in specific ways to facilitate the healing. If you have a broken arm and do not get it put in a cast by a doctor, chances are good your bone will not heal properly. The same goes for the kinds of injuries done to the brain by bullying and abuse. Like with the changing understanding and even laws around concussions, I was hopeful that with knowledge and education, teachers and coaches could change our outdated bullying paradigm that essentially is ignorant about the brain. I would not have guessed in a million years the resistance I would encounter. When I discussed Teicher's research with

Merzenich, he provided even more worrisome insight. "Teicher's documentation of changes in the myelination of the corpus callosum (like the studies that localize negative changes to the hippocampus) are examples of scientific approaches in which you see changes where you happen to be looking for them." Science, we must remember, is defined by its extreme precision and its commitment to only reporting the exact findings. Merzenich clarified that Teicher's study was either designed to examine the corpus callosum, or it identified these specific changes because they were "big enough to see." What's concerning to him is that "an enormous host of other changes might NOT have been looked for, or given the limitations of his methodological approach, not so easily seen." So while I was distressed at the idea that a crucial part of my son's brain may have been damaged, as reported in scientific research, Merzenich was opening up a far more alarming realization for me when he explained that "a change in myelination in the corpus callosum indicates a brain that is broadly undergoing 'negative' plastic revision." The final blow came when Merzenich stated: "I would bet my house that I could make thirty measures of brain function in a brain with this level of corpus callosum change, and see them ALL change, in a degrading direction."[16]

These kinds of neuroscientific findings are crucial for *all* of us who want to have healthy brains. The research is critical, but somehow it is not reaching the general public. Despite its transformative potential, it is *not* informing how teachers educate students, how coaches train athletes, how chronic disease is prevented, how mental health suffering is healed, how destructive behavior is rehabilitated, how policy is written, how workplaces are shaped. As far back as 1997, psychiatrist Stanley Greenspan was documenting the way in which only 5 percent of children suffer such intense neglect and abuse that they go on to wreak havoc with society, harming others, suffering addiction, bringing unwanted babies into the world, costing the health care and criminal justice system enormous amounts of money, and yet we do not make the changes necessary to halt this cycle.[17] Research since has shown these numbers are far higher so that by the end of adolescence, as many as 25 to 30 percent of youth have suffered "severe psychological trauma."[18] Neuroscientists have provided a vast body of knowledge over the past thirty years that should transform the ways in which we operate, but instead we remain stuck in the outdated bullying paradigm. Despite the fact that the bullying paradigm is a flimsy framework, constructed on zero reputable research, apparently it is also deeply entrenched in our brains.

In 2010, Anthes listed the kind of damage to the brain that neuroscientists were documenting from peer-to-peer bullying, and it wasn't good. In fact, it was so bad that the scientists noted that the "neurological scars" that marked a bullied brain made it resemble the brains of children who had been sexually and physically abused by adults.[19] The key takeaway of the article was that all forms of bullying and abuse, done by adults or children, were now being understood as a "serious form of childhood trauma."[20] My throat constricted. I took some deep breaths to calm down. I was arguably reading some of the most powerfully important information that I would ever come across in my lifetime. I was learning about research that was life-changing. It had the potential to completely overturn the way we thought about and handled bullying behaviors of all kinds: emotional, physical, and sexual abuse and all forms of neglect. A breathtaking scientific revolution has illuminated what bullying does to the brain, but ten years later and we are *still* conducting ourselves as if the world inside our skulls is a dark mystery beyond our control and bullying is a natural, normal behavior.

As outlined in Kuhn's influential *Structure of Scientific Revolutions*, throughout history, scientists have operated within a thought framework such as "the sun orbits the earth," "tiny amounts of mass cannot produce huge amounts of energy," or "the human brain is hardwired and cannot change." Normal scientists think within these thought paradigms and work within the rigid confines until an anomaly or malfunction appears. They strive to solve the anomaly and explain it, but nothing seems to work, and so they try harder. Then a creative thinker like Nicolaus Copernicus, Albert Einstein, or Michael Merzenich comes along. These innovators, or what Kuhn calls "extraordinary scientists," step out of the old paradigm altogether and start working within a new paradigm.[21] Copernicus puts forth the idea that the earth *actually* revolves around the sun. Einstein shares his epiphany of $E = mc^2$ to show that tiny amounts of mass can *indeed* produce a nuclear-sized amount of energy. And along with an international community of neuroscientists, and a movement that radically changed our understanding of the brain, Merzenich conducted seminal experiments showing how the brain is continuously remodeled by our experiences. The brain is "plastic" throughout our lives and not fixed or "hardwired" like most scientists, medical experts, and educators believed. *The brain is malleable and can change.* Every human individual has the innate power to change and strengthen their brain, at any time or place in life, independent from what traumas may have degraded their brain's operations.[22]

Although frequently doubted and dismissed, as well as ignored and even ridiculed for years at a time, these extraordinary new paradigms set in motion a revolution of understanding that ushered in a new way of thinking and being on the planet. Norman Doidge exclaims that the mere concept of "the 'neuroplastic brain' represents a revolutionary change in our fundamental understanding of how the brain works."[23] Nonetheless, while the revolution has taken place in neuroscientific labs across the world, we still seem hesitant to change and instead rehearse the same tired ideas that fail to stop bullying and abuse. Doidge refers to those who strive to maintain the status quo as "keepers," and he writes: "While keepers may serve science by coming up with good objections to a new paradigm, they are driven by a wish to defend the existing one, and not primarily by a quest for the truth."[24] My experience trying to change the mindset that claims we have zero tolerance for bullying and abuse, while actively enabling it, has the mark of "keepers" all over it. There are many who prefer the status quo to a quest for the truth.

Sixty years ago, we went through a comparable scientific revolution with the improved and refined quality of X-ray machines. While medical scientists likely knew there was tar in the lungs of smokers and understood the probable links of smoking-induced pathology to cancer well before the surgeon general's report, X-rays helped debunk the myths in the smoking paradigm, which was constructed on a belief (thanks to extensive advertisement) that cigarettes made you tough, cool, and glamorous. With the technological advances in the X-ray machine, experts and researchers could noninvasively document the physical status of our lungs and with these better tools discovered that smoking cigarettes actually blackened them, caused tumors, and was correlated with cancer. Likewise, in the late twentieth century, brain scans allow experts and researchers to see noninvasively into our brains. The progressively more highly resolved X-rays of the twentieth century are the progressively more definitive brain scans of the twenty-first century.

Magnetic resonance imaging (MRI) takes pictures of your brain. Functional magnetic resonance imaging (fMRI) shows your brain in movement like a video. Electroencephalography (EEG) monitors the electrical activity of your brain by tracking and recording brain wave patterns. Single photon emission computed tomography (SPECT) looks at your brain function and identifies levels of activity, as well as brain traumas or exposure to toxins and infections. Brain scans have provided evidence that bullying and abuse don't make you tough and resilient. They don't put you on the path to greatness. Bullying and abuse are in fact correlated with failure to perform, mental illness, substance abuse, aggressive behavior, weakness, chronic disease, and shortened life spans. Thirty years ago, creative scientists like Merzenich and others around the globe began constructing a new paradigm with brain imaging and neuroscientific research. It is time to enter into the new *neuroparadigm.*

Ten years ago, Anthes was hopeful in her article that—with the knowledge we had from neuroscientific studies—bullying would be seen as a "medical" problem that could be "measured by brain scans."[25] We are still nowhere near this kind of breakthrough. We are trapped in a limiting mindset, governed by a broken system, that requires us to read in mainstream media, day in and day out, about a litany of preventable abuse stories and their cover-ups. Thirty years of emotional, physical, and sexual abuse in USA Gymnastics, which has exposed similar abuse in British and German gymnastics. Rampant abuse in the most elite schools of England. Rampant abuse in the Catholic Church. Politics rife with bullying behaviors not even covered up but on public display. Rampant sexual harassment and abuse in the workplace, most spectacularly Hollywood, making the neologism #MeToo a household word. Emotional and physical abuse of athletes at Rutgers University and the University of Maryland. The "perversion files" kept on the abuse in the Boy Scouts of America and Canada. Years of emotional and sexual abuse in the Vancouver Whitecaps Football Club. Sexual harassment in the RCMP, the national police force in Canada. Elder abuse in Australia. I mean, is it just me or does this cursory glance of abuse stories from the past five years suggest that we are in need of a revolution into how we understand and handle bullying and abusive behaviors in our society?

It is abundantly clear that a bullied brain produces what I refer to as a Mind-Bully that either internally or externally attacks. It sets in motion a cycle.[26] The internal version results in the bullying continuing on "the inside," manifesting as depression, anxiety, eating disorders, self-harm, destructive relationships, substance abuse, and suicidal ideation or, at its worst, suicide. The external version results in carrying weapons, committing crimes, abusing others, aggressive behaviors, and fighting. Neuroscientists were seeing the cycle on brain scans, and it was hitting mainstream media in 2010. The research that provided visual evidence with brain images confirmed and supplied greater detail to what has been documented for decades by psychological, sociological, and psychiatric research. And yet more than a decade later, little, if anything, has changed.

In the outdated bullying paradigm, the response to disordered conduct is not to treat the traumatized brain but instead to resort to old-fashioned concepts about character and conduct so that the victims who self-harm are labeled "weak," "mentally ill," "too sensitive," or "snowflakes" and those who aggressively bully and abuse are said to be "predators," "monsters," or "snakes." These approaches are beyond outdated. They're backward. They are utterly divorced from psychological, psychiatric, and neuroscientific research. It is like treating a child who has a bacterial infection with a hot compress. It is like refusing to believe that smoking causes cancer because you can't see the blackening of the lungs with the naked eye. Helen Reiss explains: "Only when we deem all human beings worthy of respect and empathy, overcoming our natural inclinations to place them in out-groups, will civilizations reach peaceful co-existence."[27] The abuse cycle targets an often-dehumanized group into the "out-group." It's outdated because the unique brains of all human beings are the great equalizer. To exit the abuse cycle, the bullying paradigm, we must enter into a new neuroscience-informed framework that foregrounds rather than ignores the brain. Everything we use to create out-groups is constructed on what the eye can see, while what the eye cannot see, namely our brains, is dismissed, even denied. We need to get informed about our brains. Learning how to respect all brains and how to protect our brains—from bullying and abuse—is like wearing a seatbelt in your car, putting on a helmet when riding your bike, or knowing where the exits are in case of fire.

The best way to debunk a myth is with science. While we cling to an outdated set of beliefs, scientific findings continue to confirm that all forms of bullying and abuse harm minds, brains, and bodies. They do not optimize performance; they sabotage it. They don't make victims stronger; they weaken them and shorten their life span. It's time we got informed about these scientific findings. While I was laser-focused on harm being done to brains, I was not anticipating neuroscientific research that documents how remarkable the brain is at healing. The brain is vulnerable to bullying, especially between the ages of thirteen and twenty-five, but it is also unbelievably skilled at recovery. That said, if we don't know our brains are injured, it's difficult to set in motion the healing process. Most worrisome is that often victims believe that failures in their life, weaknesses, and susceptibility to addiction are their own fault. They do not know that harm has been done on the brain level and that harm can be healed.

Twenty children's and young teenagers' brains were studied by researchers at University College London using fMRI. The "f" in fMRI stands for "functional" and means that this kind of magnetic resonance imaging is done while the brain is working or functioning. This brain scan is like a video in contrast to the still image of an MRI brain imaging. All the children in the study appeared outwardly whole and healthy, but in the past, some had been physically abused and some had been emotionally abused. In this experiment, the researchers showed the children and teens pictures of angry faces, and unlike the healthy children, the abused children's "amygdalae and anterior insulae, known to be involved not only in threat detection but also in the anticipation of pain, showed heightened activity similar to that in a combat soldier."[28] These children's brains were traumatized and no longer able to distinguish a real threat from a benign one. This kind of hypervigilance takes up a lot of brain power and diverts it away from learning, creativity, problem solving, and thinking. Instead, these abused brains remain on high alert, ready for the next attack. Neuroscientists have learned that brains struggle to distinguish between bullying and abuse that is occurring and bullying and abuse that a victim is merely anticipating. Simply thinking about or worrying about abusive conduct is enough to cause the brain to go into panic mode and for cortisol to start funneling in. For victims, this can cause a chronic cycle of stress that leads to learned helplessness and does significant damage to the brain.[29] Throughout *The Bullied Brain*, we will debunk the myth of learned helplessness. With neuroplasticity, you are not helpless; in fact your potential is remarkable.

The young people in the study, who look fine on the outside, are suffering within their skulls. An amygdala on high alert, anticipating more pain and suffering, inhibits a child's or teen's ability to learn, grow, and heal. Many children with brain problems, or learning problems, are often in a state of sympathetic fight-or-flight. They feel desperate, at risk, and extremely anxious because they can't keep up with unfolding events. "The problem is that a person in fight-or-flight can't heal or learn well in this state, which makes brain change harder."[30]

Your brain needs *you* to be part of the recovery. If you maintain the belief that you are fine, that bullying and abuse toughened you up, that those who harmed you were doing it for "your own good," you might struggle to set in motion a recovery program. As we examine the harm done by bullying and abuse and then the ways to heal, it becomes clear that health resides in an integrated sense of self as a Mind-Brain-Body. When you have been bullied and abused, it can make your natural Mind-Brain-Body relationship fragment so that instead of being aligned, these parts of the self are at cross-purposes. An example would be that your mind has perfectionist tendencies that lead you to starve your body. You have an anorexic ideal of being so thin that you are putting your body and brain in harm's way. Another example would be your brain has been bullied and abused and suffers from hypervigilance. It is in such a state of constant fear and anxiety that it cannot concentrate and be productive.[31] Your mind is not consulted and is not harnessed to mindfully calm your anxiety. Your body suffers from chronic stress that research shows can lead to chronic health compromises in midlife. Once again, you are not a holistic being whereby Mind-Brain-Body are working together. There's a vast amount of research that documents how you can get that alignment back. Each chapter has a section on evidence-based practices for healing this broken relationship.

# Step 1: Harness Your Neuroplasticity

In this first step, I'm going to introduce five essential principles that will help you make an informed shift from ignoring the brain to learning about it and harnessing its neuroplastic power. Remember, first and foremost, neuroplasticity means we can change our brains.

When I present on the destructive impact bullying and abuse has on brains, inevitably a quiet line forms after the talk and people ask me with palpable concern, "Is there any way to fix the harm that has been done to the brain? Can the neurological scars be healed? Can health be restored?" The short answer is: absolutely!

That is what is so inspiring and exciting about the "Brain Plasticity Revolution," to use Merzenich's terms. Not only has he produced more than forty years of research on neuroplasticity, but he has also created programs that address specific brain traumas and distortions. Take his words to heart: "You have powers of re-strengthening, recovery, and re-normalization, even when your brain has suffered large-scale distortions that accompany developmental or psychiatric disorders, and even when it has been physically damaged in any one of the innumerable ways that can befall you in your life."[32] You have the power, or more specifically your brain has the power, to heal and restore your health. At this stage, there are some key principles that you need to keep in mind as you get informed on ways to heal your scars and recover your health if you've been bullied or abused.

First, *your brain is as unique as your fingerprints*. There is not a "one size fits all" remedy. No one can tell you and your brain what specific exercises you need to do, but there are evidence-based practices and approaches, and even highly specific training programs, that have been highly successful and that you can apply and adapt. In chapter 6, we will look at programs designed by Merzenich and his team that specifically target and heal neural networks in children and adults who have been harmed by traumas, including bullying and abuse. In the final three chapters of the book, we will look at how healing and restorative empathy, mindfulness, and exercise are for your brain.

Secondly, not only is your brain unique, it also houses at least *eighty-six billion neurons*. When scientists talk about the brain, they seek to convey just how expansive and breathtaking it is, so they mention the eighty-six billion neurons and then say that's like the number of stars in the galaxy (which isn't wholly accurate, but it's the only way to convey such magnitude).[33] Never forget that your brain is expansive. It is galaxy-sized expansive, and no one can take that away from you. If some of those glittering stars have been dimmed by bullying and abuse, they can be reignited. At all times, your brain is inside your skull just waiting for you to pull out your telescope and take a closer look.

Third, *what fires together, wires together* is a fundamental "brain change" tenet, postulated by psychologist William Hebb around seventy years ago. This capacity of the brain to change, which you read about over and over again in neuroscientific studies, is a sharp reminder that it is indeed *you* who can take charge of your brain wellness. Our society understands and has trained us to believe that if we want to get skilled at or even show talent at an activity such as music, sport, or academics, then we need to put in the hours of practice. You don't become a fabulous piano player, gymnast, or mathematician by sitting around. It makes perfect sense to us that to achieve, you must work, rehearse, train, study, fail, fall, work harder, try harder, become more refined, and seek out more sophisticated coaching or teaching. I asked Merzenich to outline what this looks like within the brain.

> Consider a practice trial in which I am trying to improve at a skill. I make a try. My brain interprets it as relatively successful. It strengthens all connections. In other words, it wires together all of those neurons that fired at each little moment in time ("fires together" moment

by moment). With each repetition, each advancing and "burning in"
change in an improving direction, you're on the path to skill mastery.

What's exciting is the way in which this brain change tenet empowers us to realize that while there's no quick fix, "extensive task repetition over an extended
period of time can result in progressive positive brain change." While with each
learning cycle, you can see improvement in your processing speed, working
memory, or physical skill acquisition on the "outside," within your brain "the
strengths of the connections between neurons, contributing to task success at
that level, grow stronger." Merzenich finishes off with "slowly but surely, your
performance abilities advance in the direction leading to performance mastery."[34]

Now, for our purposes in *The Bullied Brain*, we must realize that this brain
change tenet applies to healing and recovery. If your brain's ability to be calm
and rational has not been practiced, if that neural network has not been fired up
very often, then it's hard to wire it in. Instead, you find that due to bullying and
abuse, the neural network for anxiety, irrational thought patterns, and defensive
reactivity is your default. Any little event that throws you off means you instantly
find yourself pulled onto the default: the anxiety path of lashing out. In other
words, if your brain has been so bullied and abused that it has developed a very
defensive, aggressive, panicky neural network, where it has become hypervigilant
and flies off the handle at the least provocation, then it's because you keep firing
up and wiring in that neural network. Your brain has used those reactive behaviors to keep you safe for so long that your brain has gotten really good at it. This
angry, aggressive repeat practice has shaped your brain.

And you can change that.

You can change your brain if it defaults to anxiety, depression, or aggression
or any other neural pattern that no longer serves you. It's not easy. I don't want
you to think that there's a quick fix for this. There isn't. It takes hours and hours,
days and days, months and months of committed practice to change your neural
networks, but it's worth it. And there are ways to make this commitment happen
as we will discuss in the action steps throughout the book. With the clarity of
a scientist who has witnessed brains being redesigned and transformed for fifty
years, Merzenich believes it's important for you too to understand your brain's
ability to recover.

> While I have focused on normally ageing brains, it is important for
> you to understand that the sources of neurological difficulties and
> the principles of corrective brain plasticity apply just as much for an
> individual with a brain that is wounded, traumatized, developmen-
> tally impaired, environmentally impoverished or twisted, poisoned,
> infected, addicted, depressed, obsessed, phobic, anxious, attention-
> ally disordered, oxygen-starved, psychotic—or any one of a thousand

other brain-based maladies—as it does for ageing or, for that matter, for growing the potential of even a currently very high-performing brain.[35]

In other words, the principles of corrective brain plasticity apply to all unique brains suffering all diverse traumas. Even if you don't have a bullied brain or a brain traumatized from abuse, even if you have a very healthy, very high-performing brain, you can *still* tap into the remarkable power of corrective brain plasticity and enhance your brain performance even more. Award-winning Australian documentary maker and TV personality Todd Sampson did a three-part series, *Redesign My Brain*, where Merzenich took his high-performing brain even higher. He improved his working memory, processing speed, focus, creativity, body intelligence, visualization skills, and emotional intelligence.[36] You don't need to have a traumatized brain to optimize your performance on multiple levels.

Fourth, *your brain is like a muscle*. It can lose strength, flexibility, and endurance if you don't use it. Thus, the best way to think about changing your brain by what you practice is by seeing it as identical to how you might change your body by what you practice. Imagine that you are out of shape. You decide that you are going to prioritize fitness and get back in shape. You have to start out small. You can't go from sitting on the couch watching Netflix to running five miles. The goal is to create a reasonable plan to which you can adhere and at which you can succeed. You set daily and weekly goals. You allow your heart, lungs, and muscles to get slowly but surely stronger and able to endure more each week. It requires a great deal of mental discipline at first because the "sit on the coach" neural network has become your default pattern from practicing it so much. You've fired up the neural network for watching TV so many times that it's wired into your brain and it's what your brain wants. All of a sudden, you're changing it up and your brain feels pulled out of its comfort zone, challenged, and maybe a little curious. Tap into that curiosity. What would my life be like if I was in excellent shape?

How long it takes to get in shape varies from body to body and brain to brain. For me, it takes about six months. At the sixth-month mark, I feel like a completely different person. Instead of worrying about finding time to relax on the couch and watch a show, I find myself worrying about whether or not I'll have time to go for my run. It's a complete reversal. Your brain follows a similar pattern to your body. You can strengthen desired neural networks (like those for calmness and responsiveness) and you can make them flexible (as opposed to rigid). When you do not fire up the neural network for anxiety or aggression, they fade away from lack of use. They aren't your default anymore.

The fifth principle is that the brain has *limited cortical real estate* and so it can only retain what it deems useful to you.[37] If you do not fire up a neural network regularly, it fades away. Neuroscientists use the phrase "the brain has limited cortical real estate" to explain this principle. They remind us that on a brain level, you need to "use it or lose it." In other words, your brain eliminates neural networks that you *don't* use. If you fire up the neural network for kindness over and over again so that it wires into your brain, you no longer have neurological "real estate" for unkind or bullying behaviors. If you fire up the neural network for compassion over and over again so that it wires into your brain, you no longer have neurological "real estate" for cruelty or abusive behaviors. Imagine a society where it was required that individuals dedicate as much time to healthy social behavior—practicing the skills, firing up their neural networks—as they do to watching TV or using the Internet.

If you remove airtime or cortical real estate in your brain dedicated to the Mind-Bully, then the neural networks that support this destructive neurological network fade away. Then your restored mind, that is mindful and not mindless, can make the decision that both brain and body would benefit from a nonabusive healing session, bringing all three into alignment. As we will discuss in greater detail, research is clear that physical exercise, mindfulness, positive psychology, and dedicated brain-training exercises enhance your brain. Strengthening healthy neural networks in your brain assists your mind in making better decisions, which results in further commitment to Mind-Brain-Body health and alignment. When you establish a healthy, high-functioning Mind-Brain-Body, you are far better able to fulfill your amazing potential.

CHAPTER 2

# Refuse the Lie That Abuse Is a Necessary Evil

## *Step 2: Become a Writer of Culture*

In the mid-1990s, when Damien Chazelle was in high school in New Jersey, he was accepted into the school's very competitive jazz band where he played the drums until his hands were blistered and bloody.[1] In this band, playing the drums was infused with violence as he "tore through drumheads and drumsticks" while his ears "were always ringing."[2] Instead of feeling stage fright faced with a large audience, Chazelle recounts that his fear was focused on the conductor. "Everything that I was scared about as a drummer was him."[3] The fear that infused his days with this teacher spilled into his sleep: "the recurring nightmare of finding myself on-stage as a drummer and losing the beat."[4] Anxiety dominates his relationship to music, and it swirls around the conductor. Chazelle describes what appears to be a firsthand experience of the cult of personality in abuse. He reflects that it is "interesting when you wind up distilling all your ambitions and your goals and dreams into one single person. It's giving that person a lot of power."[5] When he looks back at his adolescence, it's apparent that he was consumed by this experience to the point of identifying with it: "I was a jazz drummer, and it was my life for a while: what I lived and breathed every day."[6]

The day-in, day-out practices. The violently aggressive teacher. It sounded so familiar to me except it wasn't basketball as with my son. It was jazz.

Ultimately, the power his high school teacher had over Chazelle as a musician and as a teenager wasn't enough to hold him, and he walked away from music to return to his earlier love of film. His time with an "abusive mentor" during his formative years led to "struggles with anxiety and depression during that period of his life."[7] The films he made after completing his BA at Harvard in visual and environmental studies are all about music. In 2014, his movie about

17

an abusive teacher of jazz, *Whiplash*, aired to critical acclaim. The protagonist of the film, drummer Andrew Neiman, is a particular target for the conductor, Terence Fletcher. Andrew practices his drums so intensely that his hands bleed in the movie.

Chazelle describes writing *Whiplash* "in a fever."[8] It wasn't about the joy of making music; it was about the "terror and the pain."[9] While a teenager, playing the drums was such a passion for Chazelle, he'd practice any moment he had for hours at a time, but ultimately his passion became entwined with the abuse he was suffering. He recalls scenes of humiliation as opposed to music. "There were so many specific things from high school jazz band that I remembered: the conductor searching out people who were out of tune, or stopping and starting me for hours in front of the band as they watched."[10] Neuroscientific research documents that for adolescents this kind of "searching out" by a humiliating teacher and being "watched" by peers is extremely stressful on their developing brains.[11]

Sometimes Chazelle sees his fear-fueled time in the jazz band as the making of him. "The end result of my personal story is that I became a really good drummer, and I know myself well enough to know that I wouldn't have without this really tough conductor and this really cutthroat hostile environment I was in."[12] Chazelle's own accomplishments make being a "really good drummer" in high school seem small by comparison. *Whiplash* was nominated for five Academy Awards and took home three Oscars, including Best Adapted Screenplay for Chazelle. Chazelle's next movie, *La La Land*, was nominated for fourteen Academy Awards and won six, including Best Director for Chazelle. He is the youngest person to ever win this award. Considering Chazelle's abusive training in music, which he then correlates here with success, you might assume that he directs his films according to the same model. However, it's just the opposite. Chazelle says that his focus with *Whiplash* was not on music; it was on the atmosphere in which he was learning and training: "It was the emotion I wanted to capture first, before the particulars of the instrument—what does it feel like to be afraid as a musician, to be afraid to go on stage, to be afraid to do something that should just bring you joy? In doing a film about that emotion, it's going to play like a thriller, because it's a movie about fear."[13] While instilling fear was the method used by his music teacher, Chazelle focuses on creating a working environment of joy and freedom when he directs Academy Award–winning films. "I like a set to be a happy place, where people can feel free to experiment."[14] Sometimes Chazelle speaks as the artistic, vulnerable teenager, but then he switches to echoing what the threatening, humiliating conductor hammered into his head at practices. "My motivation for being a good drummer was born out of fear, which, in a way, seems so antithetical to what art should be."[15] Yet he counters this belief with the one that he internalized while being a teenager ruled over by a powerful conductor: "I think there is something to be said for not coddling people and not accepting good as good enough."[16] It's as though Chazelle has two contradictory voices to portray what happened and what was learned.

What's fascinating is that returning to this experience of abuse made Chazelle scared once again, as trauma often does: "*Whiplash* scared me." But as a writer and director, he explains that he wants to draw his work "from real life," both experiences and "emotion." He wanted to ask big questions about "greatness," what it costs and what it even means. "And what does it do to the minds of young people, those who are still searching for their place in society?"[17] In leaving the powerful conductor who taught in a hostile, cutthroat environment, Chazelle taps into the courage required to return to these times of intense stress and trauma: "The more a project scares you, the more that probably means it's worth doing."[18] Chazelle has drawn us into a kind of circular logic: he learns jazz in an atmosphere ruled by fear, meant to motivate; he is motivated by that fear to walk away from music, but he circles back as a writer and director to unpack the experience of his fear in the film he wrote and directed.

After the publication of *Teaching Bullies* in 2015, I was frequently asked to give interviews, consult with families or lawyers, or present at conferences and summits on bullying and abuse. Inevitably, I would be asked to talk about the "fine line" between being tough and being abusive. There isn't a fine line, but this is a prime example of how the bullying paradigm is adept at masquerading as something that it is not. Being a tough parent, teacher, or coach, with the goal of training children to have grit and resilience, requires you to challenge them. Extensive research teaches that in an atmosphere of safety, trust, and empathy, you need to set high expectations (not impossible ones) and be committed to your belief that each child can and will achieve their personal best, especially with you supporting them every step of the way. But this is the farthest thing from being threatening, cruel, and demeaning or favoring some children and blaming others. The line between being demanding and being demeaning is as wide as an ocean.

While investigating the impact of demeaning or humiliating conduct on victims' brains, psychiatrist Bessel van der Kolk discovered that abuse victims' intense feelings of shame—that stop them from even meeting his gaze—"are reflected in abnormal brain activation." In a healthy brain, when we meet someone's eyes, the prefrontal cortex assesses the individual. The prefrontal cortex (PFC) is the last part of the brain to mature, and it's often referred to as the CEO of the brain. It's engaged in thinking about the future, weighing consequences, being rational and reasonable, and assessing a situation from a variety of perspectives. The PFC is involved in decision-making, in planning, in self-control, in social interaction, and in self-awareness.[19] Bessel van der Kolk explains that survivors of chronic trauma do not get activation in their prefrontal cortex; instead, they get intense reaction deep within their emotional brains, the "Periaqueductal

Gray." This area of the brain is involved in survival and manifests in defensive and self-protective behaviors such as startling, hypervigilance, and cowering. Individuals subjected to abuse struggle to get past their instinctive survival mode and respond to the gaze of others with confidence and curiosity.[20] If you were trained in a demanding manner (as opposed to an abusive manner), then your brain processes interactions through your prefrontal cortex. You are comfortable meeting the gaze of others. You don't startle or cower. The CEO of your brain is confident and in charge.

In 2014, a year before *Teaching Bullies* was published, I sat in the movie theater watching *Whiplash* with tears streaming down my face. It was so painful to see depicted up on the screen what I knew directly from eight student testimonies was exactly the kind of abuse Montgomery had suffered. It's one thing to hear teenagers report the abuse in short phrases and disconnected memories. It's quite another to see the abuse brought to life in a movie where you're forced to watch the intense trauma of the victims, which leads to suicide, while being simultaneously subjected to the abusive teacher's justifications.

In *Whiplash*, Terence Fletcher is a music teacher who emotionally and physically abuses his students, arguing it's necessary for them to attain artistic greatness. He looms over them, gets in their faces, exudes rage, and on occasion slaps students. He assaults them by yelling in their face a litany of mostly homophobic slurs. While the enraged yelling and misogynistic put-downs are directed at the whole jazz band, there are also specific targets. One student, Sean Casey, becomes anxious and depressed and commits suicide. His parents launch a lawsuit to try and stop Fletcher from abusing other youth. The other targeted student and the protagonist of the movie, Andrew, encouraged by his father, reports Fletcher's abuse. Fletcher is fired, but as is typical, he is hired by another band. After justifying his methods to Andrew, Fletcher ultimately asks him to play drums with his new band at a concert, assuring him he knows the music. Just before the performance, Andrew comes face to face with yet more abuse. Fletcher has given him the wrong music in another attempt to humiliate him. However, Andrew evades the teacher's final attempt to debase him and instead plays a spectacular drum solo, refusing to heed Fletcher's attempts to end the piece and stop him.

Film reviewers wonder whether the teacher is being demanding or demeaning. This confusion naturally arises in a world where bullying and abuse are normalized. In an interview, Chazelle says that Fletcher thinks: "If I have 100 students, and 99 of them are, because of my teaching, ultimately discouraged and crushed from ever pursuing this art form, but one of them becomes Charlie Parker, it was all worth it." Chazelle is quick to add that he disagrees with this "mentality" that may find talent but "causes a lot of wreckage in that pursuit."[21]

Chazelle's conflicted feelings surface again in an interview where he speaks about motivation and slides into describing abuse.

> I personally think fear is a motivator, and we shouldn't deny that. Someone like Fletcher preys on fear. I think there's a reason his methodology sometimes works, both in real life and on the screen. Fletcher's methodology is like if there was an ant on this table, and I wanted to kill it, so I used a bulldozer. Yeah, you kill the ant, but you also do a lot of other damage.[22]

Chazelle's divided feelings on the issue are shared by others trained in the bullying paradigm. In commentary and reviews, some see Fletcher as being passionate, a dedicated leader who develops talent in others, a fiery educator who wants the best out of his students. Part of the "wreckage," Sean Casey, the student who committed suicide, is dismissed. Fletcher tells the class that he was killed in a car accident, but in fact he hung himself while suffering from anxiety and depression. On his way to a performance, Andrew is also in a car accident, linking him more closely and ominously with Casey.

The bullying paradigm strives to disconnect any correlation between bullying and suicide. However, our attempts to distance bullying, harassment, abuse, and revictimization from suicide have collapsed, and we needed to invent a new word for our time: *bullycide*. The term suggests that victims cannot bear to live with the bully in their head, and victims sacrifice themselves in an attempt to escape the stress of anxiety and the hopelessness of depression. It is this kind of suicide that is at the heart of *Whiplash*, and it provokes critical questions about the bullying paradigm. Is the abusive conductor in *Whiplash* an exaggeration for the sake of making a Hollywood movie? Is a connection between emotional abuse and suicide a bit on the melodramatic side?

Ian Pace, head of performance and lecturer in music, University of London, informs us that the kind of bullying displayed by the music teacher in *Whiplash* is typical and normalized.

> What I have seen, overwhelmingly, from having gone through an elite musical training, working as a professional musician, and also from a large amount of information disclosed privately to me, is a systematic pattern of domination, cruelty, dehumanisation, bullying and emotional manipulation from unscrupulous musicians in positions of unchecked power, of which sexual abuse is one of several manifestations.[23]

In the sports world, people are quick to argue that sports are a whole different beast from other professions or training. They say you have to be *in* the sports world to understand just how tough it is, how high the stakes are, how humiliating athletes to break them is the path to glory. They say that what *would* be bullying and abuse in other situations—such as yelling, swearing, violating physical and psychological integrity, using homophobic slurs—is actually just a natural part of the passionate, intense world of competitive athletics. It's revealing to see identical abusive behavior described by Ian Pace in the *arts* world. I mean, surely a drill sergeant, a football coach, or a Boy Scout or religious leader differs in their approach from a jazz teacher or a ballet instructor. Right?

Wrong. What is shown in *Whiplash* is that the same abuse tactics that occur in the military or in coaching sports can be seen in music schools, the workplace, or Sunday School. Context is irrelevant. Abuse is abuse, but we bend ourselves into all kinds of crazy shapes to try and deny this uncomfortable reality.

In *Slate Magazine*, J. Bryan Lowder analyzes the homophobia used by music teacher Terence Fletcher in *Whiplash*.

> While the teacher's slurs modulate into anti-Semitism and other offensive realms here and there, homophobia remains his home key—which is fitting, since he feels modern culture to be irredeemably "pansy-assed." According to Fletcher, our moment's emblem—"Good job"—is an invitation to mediocrity fashioned for fags.[24]

Teenaged boys slammed with homophobia in sports are told that it's to toughen them up, but if that's the case, why does the teacher need to use it in jazz rehearsals? Do you have to be tough to play the trumpet or the piano? In our present-day silos for abuse, this would be "verbal abuse," namely, using words to hurt someone. However, this onslaught of derogatory terms is also targeted at individuals' sexual selves with the double whammy of homophobia and misogyny. It generates a feeling of fear that one might be classed as feminine, while it simultaneously slams females as pathetic, weak, poor performers, and disgusting. When young people act out these two scenarios in harmful ways—through aggression or violence to others—we set in motion the blame-shame-ostracize model without examining the relationship between how influential adults "teach" or "coach" youth and how youth behave destructively as a result.

How exactly does homophobia produce talent? I could not find any research that provided educational, psychological, or neuroscientific evidence that correlated homophobia or misogyny or anti-Semitism or racism as a way to create a talent hotbed. If we sidestep all the rationalizations for bullying and abuse in sports or music, from a neuroscience perspective, do you think the brain interprets the teacher's conduct as instructive or threatening? As noted previously, brains definitely show signs of damage on brain scans when subjected to bullying and abuse or, to use conductor Ian Pace's terms, "dehumanisation, cruelty and emotional manipulation." Moreover, as we've learned, if the brain feels threatened, it activates the fight-flight-or-freeze stress response, activating a repeated influx of cortisol, which does *not* prime an athlete or musician for top performance. As we learned, what cortisol does do is erode brain architecture on multiple levels when the stress is chronic like at repeated band rehearsals or sports practices.

Despite the entrenched belief system that abusive practices make us resilient so that we can succeed in a tough world, research is clear that abuse in all forms—from put-downs to sexual harassment—causes our stress levels to rise, which *inhibits* learning and success.[25] Why? Because, as noted before and worth repeating, chronic stress attacks the brain and damages brain architecture: "If mild stress becomes chronic, the unrelenting cascade of cortisol triggers genetic actions that begin to sever synaptic connections and cause dendrites to atrophy and cells to die; eventually, the hippocampus can end up physically shriveled, like a raisin."[26] When you learn, think, reason, and problem solve, the goal is to *make* synaptic connections in your brain; however, when you suffer from chronic stress, connections get *severed*. The goal is to grow and expand your dendrites—the branching out from your neural networks—and you also want to myelinate and thereby insulate the brain cells' transmission lines, their axons, making them more efficient. Dendritic elaboration and axonal myelination is what practice and rehearsal are all about, but when you get stressed-out, the dendrites in your brain *atrophy*. If you're in a repeat state of chronic stress, not only do you *not* myelinate, not only does the memory center in your brain shrivel, but your brain's cells *die* instead of flourishing. Abusive behaviors are described as "teaching-as-psychological warfare" and a "sustained campaign of bullying and abuse."[27] How often does it occur and simply get covered up until someone like Chazelle puts it into an award-winning movie? And even then, it's not like *Whiplash* has launched better protection for students.

Since 2013, there have been a series of NCAA coaches fired for this kind of conduct, beginning with Rutgers' basketball coach, Mike Rice, and the phrase "Jekyll and Hyde" inevitably turns up in descriptions of their personalities.[28] Likewise, psychologists use this reference to describe split-personality in domestic abuse: "unmasking the terror of Dr. Jekyll and Mr. Hyde."[29] These terms are taken from a nineteenth-century novel written by Robert Louis Stevenson. In the story, there is an upstanding, respected, caring doctor, Dr. Jekyll, who has an alter ego who is violent, destructive, and callous. He takes a potion he makes that changes his appearance into someone else, Mr. Hyde, so that he can indulge himself without detection. He loses control of his ability to put on and off the monstrous, murderous Mr. Hyde and eventually commits suicide. Even his closest friend and his family do not realize that Jekyll and Hyde are the same person. Likewise, in abuse cases, people are routinely amazed that one single individual can house two totally different personalities: one, a respected professional, well-educated, refined, and popular like Dr. Jekyll; and two, a violent, obsessive, harmful abuser like Mr. Hyde.

*Whiplash* accurately depicts the Jekyll-and-Hyde split personality of an abusive individual who is able to switch from abuse to charm with the bat of an eye. Fletcher is described aptly as "a seductive monster, swiveling from charm to nonchalance to violent rage with a snap of the fingers."[30] Note that "seductive monster" is a dehumanizing term unto itself and serves to convey how both victims and perpetrators become dehumanized in the bullying paradigm. Neither are discussed in terms of having damaged or disordered brains. Neither undergo brain rehabilitation or brain training to recover and heal. Merzenich refers to a child, whose brain is harmed or distorted in childhood so that he or she grows up bullying and becomes abusive, harming and distorting victims as a "mutual tragedy." Despite the vast amount of scientific research that informs his understanding, he puts it simply: there is something "neurologically amiss" in the bullying and abusive brain. Individuals inadvertently "learn" the behavior; their brains are repeatedly stimulated by bully-mentors. Being abused in childhood is the way that "weaknesses and distortions" in the brain that lead to "destructive and manipulative behaviors" are engendered.[31] Trying to assign blame, from a neuroscientific point of view, is like trying to figure out which came first, the chicken or the egg. Maybe we could take the time we spend trying to assign blame and instead focus it on healing the damage done and stopping the cycle. According to research, the effects of being bullied "are direct, pleiotropic, and long-lasting, with the worst effects for those who are both victims and bullies." While those who escaped bullying in childhood are safe, those who were victims, or worse were victims who then became bullies, are "at risk for young adult psychiatric disorders."[32] It is a *mutual* tragedy. All brains involved need healing.

Research into bullying and abuse reveal four traits referred to as the "Dark Tetrad" by experts. The four overlapping personality traits of one who bullies and abuses are: narcissism, psychopathy, sadism, and a manipulative deceptiveness known as Machiavellianism.[33] The narcissist is so self-focused that he doesn't have room or space to see beyond his own reflection and register the pain or suffering of a victim.[34] Psychiatrist Stanley Greenspan notes that many people who are emotionally "blocked" due to emotionally abusive or neglectful adults in childhood "inflict grievous, even fatal, harm on others, quite unaware that their victims have feelings every bit as strong and valid as their own."[35] Narcissists almost always come by their cruel inability to see their conduct as harmful and their belief that *they* are the victim from childhood suffering that blocks natural, emotional development. Narcissism is a "serious disorder characterized by lack of empathy, grandiosity and impaired emotional regulation," which is associated with "brain irregularities."[36] It's a serious illness that manifests in a grandiose sense of self who is prone to lying, who can be charming and clever, and who resorts to cunning and highly manipulative behavior.

How does narcissism appear inside the brain? While a brain that has *not* suffered abuse shows activation during empathy, those who have developed narcissism show "significantly decreased deactivation during empathy, especially in the right anterior insula." Furthermore, neuroimaging data "indicates lower activity in the insula in high narcissistic subjects."[37] A recent study showed that patients who had brain lesions due to removing tumors in the anterior insular cortex (AIC) showed deficits in empathic pain processing. What's eye-opening to learn is that "empathy deficits in patients with brain damage to the AIC are surprisingly similar to the empathy deficits found in several psychiatric diseases" including borderline personality disorder and narcissistic personality disorder.[38] If you learned that someone was behaving in destructive ways because they had lesions on their brain from the removal of a tumor, would you feel empathy for them? What about if they had comparable lesions on their brain from childhood abuse? We tend to think about the former as a medical issue and the latter, which can manifest in highly destructive behavior, as a conscious choice and a sign of poor character. The goal is not to excuse this behavior—just the opposite. The goal is to see this conduct as infectious, needing quarantine and rehabilitation. When people are suffering from a medical crisis such as a tumor or lesions, we strive to heal them with our most up-to-date approaches. Why is it not the same when they are suffering from comparable brain damage due to abuse?

Those who demonstrated more intense forms of narcissism also showed "lower connectivity between certain brain areas, including the prefrontal cortex and ventral striatum." When these areas are well connected and integrated, it allows a person to think positively about herself; whereas low activation in these brain regions may push a person to seek out affirmation from others. This profile fits with psychological studies of narcissists, where beneath a confident and self-aggrandizing exterior, they actually suffer from low self-esteem and self-loathing. Their lack of self-concept makes this difficult, perhaps even impossible for them to understand.[39] While the neuroscience of narcissism is far more complex, the key point at this juncture is to realize that the brain has been affected and the brain needs healing.

While sadism and psychopathy are intensifications and overlapping traits of narcissism, Machiavellianism has some distinct features. In research, when compared to control groups, Machiavellians lack social-emotional understanding, empathy, and emotional intelligence. They "strike first without prior provocation," manipulate others for "short-term gain," and are characterized as having strong desires to succeed frustrated by "low mental, social, and emotional abilities." The tension between high expectations for themselves and those expectations continually being frustrated leads to their "indication of disgust towards others" and their compulsive and fanatic attempts to "win at all costs" and "outsmart" others.[40] Machiavellians "look for situations that allow them to cheat and not get caught," and as a way to cover up for their mediocrity, they "ingratiate" themselves with those in administrative, managerial, or leadership positions. In a study of Machiavellians, researchers found "significant positive differences" in parts of the brain involved in seeking rewards, regulating negative feelings, planning, outsmarting people, feeling disgust, and suppressing negative frustration. The researchers note that "all these areas are related to Machiavellian tendencies such as engaging in sneaky political maneuvering."[41] With even an introductory overview of a Machiavellian brain, it makes it easier to understand how our system struggles to identify and manage these kinds of individuals. Those in charge may well be as manipulated as those who are victims of these kinds of brains.

Leaders and victims become confused and believe the manipulations of a Machiavellian or the claim to innocence of a narcissist. This is why they are referred to in expert literature as "snakes in suits."[42] Using the term "snakes" is dehumanizing, of course, but leading researchers Paul Babiak and Robert Hare note that the defining trait of these highly destructive individuals is that they are "without conscience," at the same time as being "highly skilled communicators in the language of subterfuge, deflection and threats."[43] This is where the traits of sadism and psychopathy further define the Dark Tetrad. Since biblical times, the snake is the symbolic figure for manipulation, making you believe that dark is light, promising greatness while destroying it, offering power while stripping the victim of selfhood. Reading Babiak and Hare's work is as frightening as watching a snake slither into an unsuspecting classroom, gym, place of worship, home, or workplace. How might we better handle these kinds of destructive individuals if we were informed about the brain distortions under which they struggle and suffer? Again, the goal is not to forgive and forget, dismiss and deny. The goal is to seek insight, diagnosis, rehabilitation if possible, and most importantly protection for victims.

While Chazelle focuses on the impact Fletcher has on his music students, Martin Teicher used technology to examine the brain of an abusive teacher, and he discovered that what makes someone become a "Jekyll and Hyde" is childhood abuse. It is a cycle. The medical term that describes a Dr. Jekyll and Mr. Hyde is "associative identity disorder." It is when two "seemingly separate people occupy the same body at different times, each with no knowledge of the other." Keeping this in mind, perhaps it is possible that at times the abusive individual truly does believe they are innocent. Teicher says "associative identity disorder" is a more severe form of "borderline personality disorder."[44]

Borderline personality disorder is another way to understand those who bully and abuse in very targeted ways, while appearing charismatic, popular, and well-regulated in other circumstances. In a sense, their brains are "broken" into two distinct selves. These "abusive narcissists" have "one dramatically changeable personality with an intact memory."[45] If they were a fully split personality, they would not cover up their abuse because they could not even remember that they had done it, whereas most often, abusive individuals work carefully to keep the abusive conduct hidden. If you read an article in the media about an abusive individual, it is almost guaranteed that people will rise to the abuser's defense and tell anecdotal stories about how wonderful, kind, and caring they are. This casts doubt upon those who report the abuse. However, from a neuroscientific point of view, it makes perfect sense that the individual, most likely once a victim, has developed a brain disorder—a split personality—that targets certain individuals, offers kindness and privileges to others, ingratiates themselves with those in power, covers up their abuse, and thereby perpetuates the abuse.

In the late 1990s, neuroscientists were studying bullying behaviors or "social subjugation" in animals and examining what was happening in their brains. While documenting the harm done to the victims, their findings also indicated that those harmed were more "likely to become bullies themselves." We don't know what happened to Fletcher while he was growing up or during the most vulnerable time of adolescent brain development, but we can reasonably guess he was once a victim. Juvenile animals who "were bullied by an adult became more aggressive toward smaller animals, while being fearful and subordinate" with equals and those who were larger.[46] Once again we see the cycle: a bully victim grows up to be a victimizer. In other words, when adults are abusive to children, you'll find children bullying. Put another way: where you have violence against children, you'll find children growing up perpetuating that violence. The results of this study suggest adolescence may be a sensitive period for the development of aggressive behavior in adulthood, leaving film audiences wondering what happened to Fletcher during his childhood and teen years.

Chazelle's film is compared to sports and war movie plots: the abusive "drill sergeant" or coach who breaks down the individual until he or she rises from the ashes like a phoenix to triumph in a final competition or battle. Andrew has a breakdown under the abuse, and with the encouragement of his father, he reports Fletcher, who is then fired. In the final scene, the abusive teacher grasps at a final opportunity to publicly humiliate Andrew by giving him the wrong music at a performance. This final attempt at humiliation differs from sport and war plots where there is more often a meeting of minds or reunion of sorts between the drill sergeant and soldier or the coach and athlete.

Andrew avoids his abusive teacher's final attempt at public humiliation and instead gives a brilliant, untouchable drum solo. Ironically, those who have been indoctrinated to believe students should be sacrificed to sport wins or art performances use this moment as support for the bullying paradigm. They see this scene as proof that bullying and abuse *are* indeed a necessary evil to create and achieve greatness. Fletcher certainly argues that if you want to achieve beyond mediocrity, the place for "fags" and "pussies," then you must submit to bullying and abuse. However, Chazelle did not name his film with a musical title that even hints at triumph. He called the movie *Whiplash* after a song by Hank Levy, but the term describes an injury that is invisible. A "whiplash" is a closed head injury that occurs within the skull and may well harm the brain. Those who see the student, Andrew, as whole and successful at the movie's end are unable to see that he might have a fragmented or "broken" brain from the abuse he has endured. This is why the scene where he denies depth or history to his girlfriend and then dismisses her from his life is so telling. He is not, according to Chazelle, becoming a mini-Fletcher who discovers talent. No, he's becoming the mini-Fletcher who uses others as mirrors on which to project his own self-dismissal and self-loathing. As Chazelle puts it, he becomes "unforgivably cruel."[47]

Journalist David Sims calls Andrew's final performance a "Pyrrhic" victory to convey that his win inflicts such a devastating toll that it might as well be a defeat.[48] Chazelle, who knows teacher abuse on a deeply personal level, did *not* triumph from what his music teacher did to him. Chazelle suffered, like so many students today, from anxiety and depression. Like a whiplash injury or a concussion, anxiety and depression are serious wounds, but they are invisible without brain-imaging technology. Chazelle has Andrew bleed when he practices in an unhealthy, demented way to please his abuser. Furthermore, he puts at the heart of his film the suicide of the teacher's other target, Sean Casey, a stark reminder to the audience of the thirteen students a day who kill themselves in the United States alone. Any kind of triumph, in music or performance or band cohesion in the final scene of the movie, cannot be discussed without reference to one of the teacher's abuse victims committing *bullycide*, while the other, who appears on the outside as successful, is actually bleeding and suffering from invisible damage like that sustained in a car accident. We are so well-schooled in the bullying and abuse paradigm that we only see the success. We need a movie to show us the blood.

Even though Chazelle's abuse occurred in his public high school in Princeton, New Jersey, the music school in *Whiplash* is often discussed as a fictional Julliard School of Music in New York City. In the film, scenes in the band room were shot in Julliard. After *Whiplash* screened, professional musician and jazz faculty member at Julliard Michael Sherman was interviewed about abusive teaching practices at the renowned school of music. Sherman does not mince words: "This type of mental and verbal abuse, bordering on physical, is taken so seriously that he'd be thrown out of Juilliard and most schools no matter how great he was."[49] Now, here is the worrisome part. This is Sherman's final comment on the film in the same interview.

> I actually cried at the end, when the kid was kicking ass in the last
> tune. And the most important thing about that scene is maybe what
> the band director was trying to get out of all the students from the
> very beginning—my take on it, at least—which is he's trying to teach
> them to be leaders like he is. And at the very end, [Andrew] says, "I'll
> cue you." He's the leader now.[50]

Do you see what's happened? It's amazing. On the one hand, emotional abuse
is *not* tolerated at Julliard or at any school. On the other hand, when a teacher
abuses students, it's not actually abuse; it's leadership training: "He's trying to
teach them to be leaders like he is." This is just a stunning example of the bullying paradigm that has infiltrated our brains. Sherman is like a split personality.
One part of himself is very clear and truly believes abuse would not be tolerated,
but then another self emerges who states just as clearly that what looked like
abuse was actually an exercise in leadership and the lucky victim has learned his
lesson and "he's the leader now." If Andrew has become anything in *Whiplash*,
it's quite possible he's becoming an abusive narcissist or, as Chazelle puts it,
"unforgivably cruel."

Sherman's comment is concerning because it is a glaring example of the way
in which abuse is normalized. It exposes how quickly we forget abuse when we
are distracted by success or performance. How does Fletcher role-model leadership in his final attempt to humiliate his student by giving him the wrong music?
Is it possible that when the student, who has escaped the abusive teacher and
reported him, says, "I'll cue you," he's expressing his freedom and independence
from the psychological trap constructed by his teacher? If so, this would mean
that the excellence in his drum solo signifies his triumph over abuse and his own
creative style, his own fulfilling of potential. It is shocking but also typical that
Professor Sherman, along with others, sees the student's success as attributable
to the abusive teacher.

This way of thinking manifests in many other fields. The football team or
softball team wins a game, a streak of games, or a championship, and we forget that student-athletes were harmed or even died. We walk around clapping
people on the back saying, "No pain, no gain," or we say the team sure came
through the regime of "blood, sweat and tears." The corporate manager, who
has multiple victims reporting his abuse, lands a big client and the abuse is seen
as a necessary evil to serve the bottom line. The abuse of film producer Harvey
Weinstein is swept under the carpet as he does yet another fabulous movie. The
fictional drummer, who we have watched suffer intensive abuse throughout
*Whiplash*, plays a fabulous drum solo, and that is enough for Sherman, a real-life
teacher at Julliard, to forget about the abuse.

And let's take it one step further: Sherman has turned the abuse into a quest
for leadership and believes that the student Andrew is becoming "a leader,"

not an abuser. Hope you see this as gaslighting. In *Whiplash*, Chazelle shows the way in which Andrew is turning into his abusive teacher: he's "becoming a mini-Fletcher" in the way he treats his girlfriend. As Chazelle puts it, Andrew "purports that he can just sit down and X-ray into her."[51] Chazelle clarifies that he learns this kind of narcissistic approach from Fletcher. He explains the pattern he developed in *Whiplash* to show Fletcher and "mini-Fletcher." He says in an interview, "Terence Fletcher has one conversation with Andrew and then decides he knows his entire backstory, and knows exactly how to twist the knife."[52] While Chazelle himself clearly escaped the fate of passing on the baton of abuse from leader to follower, abuser to victim, he depicts it powerfully on-screen.

When Rutgers basketball coach Mike Rice had to resign after being exposed in the media as physically and verbally abusing players, his assistant coach, Jimmy Martelli, instantly resigned as well. The players referred to Martelli as "baby Rice" because he mirrored the abuse of Rice.[53] Dating back to the time of Sigmund Freud and still being discussed in recent neuroscientific studies, a witness to repeated bullying and abuse may well normalize the behavior or mimic it: "As a psychological defense mechanism, the child may convince herself that other people deserve to be treated unkindly or thoughtlessly, and begin to copy the same sort of behavior."[54] The targeted victim does not ever *deserve* abuse. And ironically, studies show that in particular, emotional abuse—belittling, humiliating, shouting, scapegoating, rejecting, isolating, threatening, and ignoring—is more likely to increase when the target is identified as talented.[55] In other words, abusive individuals target talent, but those who mimic them have fallen so deeply into the abusive belief system that they think the fault lies with the victim who *deserves* mistreatment for some kind of failing, fault, or lack.

Passive witnesses or active imitators—who identify with the aggressor—are frequently called upon when multiple abuse reports take place and an investigation or media exposure launches. These individuals report that there was *no* bullying or abuse that happened, and in fact, the producer, manager, teacher, or coach in question was *not* cruel or abusive. Actually, they were caring and supportive. Actually, they will say, the victim *deserved* the treatment because they were weak and lacked talent. It was their fault. Knowing the psychological and neuroscientific reasons why bullying and abuse witnesses or imitators speak this way, surely we can recognize that if someone identifies with the aggressor, it's a defense mechanism and survival strategy. While they come by it honestly, their perpetuation of the bullying paradigm puts them, and society at large, on an endless treadmill.

That said, *Whiplash* is a story told by Chazelle, who turned his experience of being a teenager exposed to an abusive music teacher into a tale of triumphing over trauma. He reversed the narrative that says victims who commit *bullycide* are just weak, that they suffer from mental illness, that it's their own fault.

Instead, frame by frame, Chazelle unpacks Fletcher's deadly combination of power, violence, and humiliation that breaks an individual asunder until he can no longer experience himself as one aligned whole. It seems that Sean Casey has fallen into borderline personality disorder, whereby the internalized bully, the Mind-Bully, so loathes and lashes out at the self who wants to study music that he destroys him. Alternatively, the self who wants to study music is so sick of the bully in his head that he is willing to sacrifice himself to stop the abuse. Chazelle exposes, throughout his award-winning film, that the abusive teacher is seductive, manipulative, and quick to justify his abuse as necessary to create talent. However, writer and director Chazelle doesn't fall for the bankrupt notion and blows it up in his movie. He refuses to be a reader of culture, an adherent to the bullying paradigm, and instead charts a new course as a writer of culture. The great power that Chazelle had to walk away from negative forces, ones that try to harm you and pull you down, is a belief in himself. His high school jazz teacher failed in his quest to bulldoze Chazelle's belief that he was going to be creative and make movies.

Chazelle doesn't pin his remarkable success on being abusive; he attributes it to his willpower. What fuels his willpower so that he stays the course, has the discipline, works so incredibly hard? Listen to what he says: "If you want to make a movie, there may be many forces trying to pull you down, but really, a lot of it is will power. You can will it into being if you just believe that you are going to make a movie."[56] We will look more closely at this concept of self-belief in following chapters, and we will also look more closely at those who stop believing in themselves. This is not their fault and is simply heart-breaking.

# Step 2: Become a Writer of Culture

The action plan that accompanies this chapter is a challenging one. It requires you to think independently of your social and cultural training. This kind of critical thinking is truly hard work but very good for your brain. The more you dispel the clouds of false thinking and replace them with clarity, the healthier, more flexible, and more resilient your brain becomes. If you have found yourself confused, emotionally overwhelmed, suffering from cognitive dissonance due to the mixed messages in our society about bullying and abuse, you are not alone. In fact, high levels of mental illness in our society may well correlate with the conflicting messages we get about abusive behavior.

Because abuse is a cycle, those who now occupy positions of power may well have brains so steeped in normalized bullying and abuse, so shaped by a culture constructed on bullying and abuse, that they can no longer think clearly.[57] The main takeaway from this chapter is that, if you want to exit the hypocrisy of the bullying paradigm and shield your brain from the onslaught of mixed messages, then you must stop being a passive reader of culture and become an active writer of culture. Chazelle became a writer of culture after having been in an abusive environment. You do not need to share your writing with anyone but yourself, but the practice of rejecting the myths, falsehoods, and outright manipulations that prop up the bullying paradigm is a good early step in healing scars and restoring health. Becoming a writer of culture can take many forms. You can keep a journal, make recordings, take photographs, paint, draw, cartoon, collage, or create a film.

Just by reading this book, you are at the vanguard of a dual revolution: first, learning about your brain and, second, exposing the myths of the bullying paradigm. At the outset, you are asking questions about why we are taught next to nothing about our brains. You are wondering why teachers are not educated in how brains learn and why mental health practitioners are not trained to work with brain images. You want to figure out how it is that coaches know so little about brains when sports are a "mental game." You are becoming curious about why workplaces do not have even basic knowledge about what leads brains to perform at their best and what inhibits brain optimization, despite the reality that we are now in a "Brain Economy."[58]

Secondly, you are discovering that what we are taught about bullying and abuse in school, on sport fields, in arts programs, in governance and politics, and at the workplace is ignorant at best and outright harmful at worst. Children are told not to bully one another, yet the authorities in their world look the other way when parents, teachers, religious figures, or coaches are abusive. When children bully, it's unacceptable. Yet when adults bully, oftentimes child victims are blamed. Children learn bullying behavior *from* adults, but no one talks about this transfer of destructive behavior. Children are made to feel that somehow the fault lies with them. They were born into the world, not "trailing clouds of glory," like poet William Wordsworth says, but with a bullying agenda, apparently meant to distress established adult society. It's ridiculous when you think about it and yet so hard to see when we've all been trained to think this is normal.

The takeaway from this chapter and the action step I recommend is for you to write an alternative plan for yourself. Imagine and articulate a different paradigm: a framework that does not contradict, one that is not hypocritical, one that halts the cycle. What would happen if you exited the bullying paradigm and became a writer of a new culture, one informed by neuroplasticity? If you can change your brain, if you can heal your brain and make it stronger, then you can bring the abuse cycle to an end. It's certainly worth a try.

If bullying and abuse cause splits in ourselves so that we develop a dual personality—as simple as one that tells children not to bully and then turns around and bullies children and adults, saying it's for their own good—we need to consciously work to mend these rifts in ourselves. Take some time to let go of any labels that might have been applied to you in the broken system ruled by the bullying paradigm. Try to side-step binary oppositions that force you into one camp or the other: instead of seeing yourself as a victim, explore how you were put in that role. Instead of seeing yourself as a perpetrator, look farther into the past to explore how you might have used abusive behaviors as a survival strategy. Is it possible you've developed a Mind-Bully or have outwardly used bullying behaviors because you've identified with the aggressor? Have you been a bystander? Did you take on the whistleblower role? We cannot exit the bullying framework without being as reflective as possible about our own roles in it over a life span. Few of us have not been infected with bullying and abuse. The question is: what do we do with our various traumas? We have brains that can heal and recover so it's time we took some proactive steps away from the bullying paradigm and toward something new. The new paradigm is going to be as unique as your own brain and your own story. This action step is about envisioning an exit from the bullying paradigm and an entry into something you have designed that is more likely to heal your scars and restore your health.

Replace the labeled fragments with a more holistic story of your Mind-Brain-Body. How might they offer a more unified approach to your life and better shape your agency? You are not a victim in the present because it is within your power to change your Mind-Brain-Body. Whatever might have happened in the past, when you were a reader of culture, is done. It's over with. There is nothing you can do about the past, but you have incredible powers within your brain to change your present and future. Imagine writing a present and future for yourself in the new paradigm whereby your Mind-Brain-Body are aligned and working together for your health and happiness and the health and happiness of others. Once you believe in yourself, you can put in the deep practice, the hard work to fulfill your wondrous potential.

Chazelle is clear that a teacher like Fletcher is not the one who finds or creates a "Charlie Parker." After being abused as a teenager, after rebelling against the bullying paradigm and all its justifications in *Whiplash*, Chazelle states unequivocally that abuse is *not* a necessary evil to achieve greatness. What does he believe is the crucial ingredient to rising above mediocrity and attaining art? In one word: work.

> I don't think you have to suffer for your art. That's just a romantic notion that I think is not always right. At the end of the day the only common thread that geniuses in any discipline have is that they work harder than anybody else and this notion that they just roll out of bed born with it is just utter horse shit.[59]

# CHAPTER 3

# Learn to Disobey

## *Step 3: Grow Your Talent*

In 1963, Stanley Milgram, a psychologist at Yale University, wanted to explore how it was possible that ordinary German citizens carried out heinous acts against their fellow citizens during the Holocaust.[1] Many of these citizens, accused of war crimes in the Nuremberg Trials, argued that they did *not* have responsibility for what they did, namely cruelty beyond imagining and mass murder, because they were merely following orders. This argument surfaces frequently in the bullying paradigm whereby those accused of abuse proclaim innocence to the point where they convince themselves—and others—that *they* are actually the victims. As noted, it's remarkable how quickly accusations of abuse, from multiple victims, transform into "This is a witch hunt." The reversal is surprisingly swift, from speaking up about abuse to finding yourself in a victim-blaming scenario. The reversal turns abusive acts inside-out so that there is never any accountability. The Holocaust was carried out by a million Germans, but the millions of deaths were supposedly the fault of a handful of authorities pulling the strings.

Milgram decided to test whether the average Joe would do terrible things if commanded by someone in authority. Not surprisingly, he chose an educational context to test his theory. In the experiment set up by Milgram at Yale University, there were three roles: the "authority figure" (played by an actor), the "teacher" (a random person who had agreed to participate in the experiment), and the "student" (who was played by Milgram himself or another person who was in on the experiment).[2] The random person, the average Joe, who agreed to participate in a research study for Yale was asked to administer electric shocks to a person who was behind a glass wall and was expected to answer questions.

Behind the glass, this student (played by Milgram) was only given an electric shock if he got the answer wrong. The average Joe, playing a kind of teacher role, had a dial with visible markings on it that showed just how much electric shock was being sent to jolt the student. Standing by the average Joe throughout the experiment was the authority figure, who was dressed in a white lab coat and appeared for all intents and purposes to be a serious Yale University researcher, in other words, not someone you'd question lightly. The random person, the average Joe, had no idea he was working with an actor as authority figure and Milgram himself as the student that he was being asked to shock. The electrical jolts the average Joe was expected to administer, and that he believed were real, in fact were fake. He was simply told that the student needed to answer questions correctly or be penalized by an electric shock. The shocks were on a scale from 15 volts (slight shock) to 450 volts (severe shock). The scale showed clearly that at 300 volts, the electrical jolts were hitting a threshold of danger to the student.

The experiment was repeated forty times with forty different participants. At a certain point, the student—remember he was being played by Milgram—would begin to show signs of suffering that would increase to the point of writhing and screaming in agony. Milgram was convincing in his role as student being shocked with electricity. If the average Joe, in the role of teacher, resisted the authoritative researcher's order to increase the voltage, the authority figure, in his white lab coat, would prompt the average Joe according to the following statements, always given in order:

**Prod 1:** Please continue.

**Prod 2:** The experiment requires you to continue.

**Prod 3:** It is absolutely essential that you continue.

**Prod 4:** You have no other choice but to continue.

These are oppressive commands. The average Joe is under pressure for sure, but remember: he's just some random guy off the street who has agreed to participate in a research study. It's not a high-stakes situation if you think about it. The average Joe is a regular citizen who thinks he's helping out some professor at Yale University. It's not like he *has* to follow through or can't say to himself, "Forget it; I didn't sign up for this. These guys are crazy. I'm not going to hurt this person behind the glass wall just so some Yale prof can publish a paper about the results and get tenure. I'm out of here!" The average Joe could have said all of these things and walked right out the door. No one, in fact, is forcing him to shock some hapless student who can't get the answers right.

Now, before we discuss what Milgram discovered, before we talk about the results of this experiment, I want you to test your own assumptions about obedience to authority by answering two questions.

**Question 1:** From zero to forty, how many individuals do you think electrocuted the student almost to the danger level of 300 volts?

Record your answer for further reflection.

**Question 2:** From zero to forty, how many individuals do you think electrocuted the student at a level of almost 450 volts?

Record your answer for further reflection.

Here is what Milgram discovered.

**Answer 1:** *Every single one of the average Joes* in the role of teacher electrocuted the student, at the command of the authority figure, up to the dangerous level of 300 volts.

Just sit with that for a moment. What does it tell you about citizens and compliance? What does it tell you about how much we hammer into children that they must obey so that they grow up to be adults who obey? What does it tell you about the bullying paradigm, within which we appear to be raised and educated so that harming someone else seems to be fine as long as you're *instructed* to do it? What this should tell you is that if your brain has had grooved into it the bullying and abuse mantra, just like a vinyl record going around and around, it's hard to simply pull up the needle, remove that record, and put on a new one. It might not be easy, but it *can* be done. The average Joes in Milgram's experiment have neuroplasticity like the rest of us, and they could go into an intensive program of rehab that teaches them day after day that shocking people with dangerous jolts of electricity is actually *not* okay. They don't have to simply do as they are told. They can get up, brush themselves off, and walk right out of the cage of learned helplessness. They're not helpless. They can resist. They can refuse. They can rebel against the authority figure. And so can you.

What I find particularly chilling in Milgram's experiment is that there aren't any real risks. The stakes are super-low for the average Joes, and they *still* simply obey the directive to hurt someone and endanger them. Compare that to actual situations, when those who are bullying and abusing under the guise of authority often have the power to issue significant threats and follow through on them. Then you're dealing with a serious dilemma: resist and risk my position, my social standing, my letter of reference, my opportunities, my scholarship? Resist and risk my job? Resist and risk my family? These are frequently the bars in the cage that keep us quiet and compliant when we're dealing with those who bully and abuse. The average Joes didn't have any of these risks and *still* they obeyed.

So with that depressing thought, let us look at answer 2. Remember, the question was: From zero to forty, how many individuals do you think electrocuted the student *beyond* danger to severe shock at 450 volts?

**Answer 2:** 65 percent of the average Joes in their role as teachers electro-cuted the student, prompted by the authority figure, up to probably fatal shock levels of 450 volts.

It's incredible when you think about it, but that's just it. Our brains have been trained *not* to think, not to feel, not to question. We are trained from early childhood to comply. We are trained to obey. Inside the head of the average Joe, there was clearly a Mind-Bully at work telling him that he had to do as he was told, telling him that he had no authority to speak up and refuse to do what was asked. Who did he think he was anyway? Those who bully, and those who then internalize the aggressor as the Mind-Bully, love to use rhetorical questions that stress the victim's worthless, voiceless, powerless position. In basketball practices, one teacher would stop the play and point to the floor in front of him and Montgomery would have to stand there. Then the teacher would yell rhe-torical questions in his face: "What do you think you're doing? Do you even like basketball? Do you even deserve to play?" The questions seem innocent enough, but they were yelled inches from his face. Players described it as "vicious." They said they wanted to stop it. No one did though because he was the "teacher." The other teacher watched, ensuring the scenes of humiliation were normalized. Montgomery would hit a point where he'd try to get away and the teacher would grab his arm or jersey and keep yelling rhetorical questions. I asked Montgomery how often this happened. His reply: "hundreds of times." He never answered these questions because he was paralyzed with shame and fear. When you hu-miliate a teen in front of their peers, it's arguably the worst thing you can do to their developing brain. Montgomery never came home and told us what was happening. We didn't learn until the other boys on the team began giving their testimonies. The Mind-Bully that tells you to cast your eyes down and remain still and silent (freeze) is trying to protect you. The Mind-Bully that tells you to run (flee) is trying to protect you. And the Mind-Bully who lashes out at the ag-gressor (fight) is also trying to protect you. When you imagine that kind of brain training, you can empathize with the average Joes. Faced with the training to freeze, flee, or fight, they froze and did as they were told. Why do bystanders to bullying and abuse stand by and not intervene? They've been trained to stand by. Their Mind-Bully knows it's risky, perhaps dangerous, to try and stop the harm.

The average Joe's Mind-Bully would have misinterpreted the researcher as a risk, a threat to survival, if the average Joe dared to assert himself and exit the cage. The Mind-Bully figured—in its role as defender or protector—that the average Joe was in a certain amount of danger, and it would be better to simply follow orders than get into any kind of trouble. Sadly, the Mind-Bully of 65 percent of the average Joes must have had quite a lot of experience with the perils of speaking up, resisting, and rebelling against authority. Therefore, even in a research experiment, the Mind-Bully believed keeping quiet and obedient was the safest course of action. Remember Bessel van der Kolk's research that showed abused individuals were hypervigilant; they cowered and startled. They couldn't meet the authority's eyes. They were processing the situation from the emotional center of the brain, not from the prefrontal cortex, which is where we make rational, calm, moral, thoughtful decisions. What's notable in Milgram's research is these kinds of individuals are not outliers; they're the norm. Milgram discovered that "people obey either out of fear or out of a desire to appear cooperative—even when acting against their own better judgment and desires."[3]

After the experiment, the average Joes were told that there weren't any real electric shocks. One of the average Joes broke down in tears because he thought he had killed the student and was deeply relieved. Milgram divided the responses of the average Joes into three categories. The first category was made up of those who obeyed the authority but justified their actions. This is the textbook position of those who bully and abuse. They dismiss their actions. When that doesn't work, they deny them. They are always innocent. They are always victims. They assign blame elsewhere. Milgram found that this category of Average Joes blamed the experimenter or the learner: "He was so stupid and stubborn he deserved to be shocked."[4] The term "deserved" is a key word for those who bully and abuse. The victim deserved it. They were taught a lesson. The second category of average Joes blamed themselves. They felt guilty about what they had done and were hard on themselves. They probably didn't internalize the aggressor and may even be more inclined to question authority in the future.

The third category was the rebels. "They questioned the authority of the experimenter and argued there was a greater ethical imperative calling for the protection of the learner over the needs of the experimenter."[5] This is the category where you would find extraordinary scientists like Marion Diamond, Eric Kandel, and Michael Merzenich, the ones who dare to question the established paradigm and begin thinking and working within a new framework. This is the category Montgomery would have been in, along with Damien Chazelle; however, they could not rebel until they walked away from the abusive individual(s). Why? Because they were children in an abusive regime and survival comes first. Speaking up, especially in childhood, is one of the hardest acts a person can do. It requires immense courage and sacrifice. Montgomery knew the risk and ended up losing competitive basketball. Chazelle lost playing jazz drums. These were passions that the bullying paradigm forced them to lose because it condones and enables abuse, even when children report and seek protection. Milgram's experiment was designed to find out why only a minority would challenge authority, and he discovered that so "entrenched is obedience it may void personal codes of conduct."[6] In other words, individuals have been trained in obedience or compliance to the point where it overrides their personal ethics or integrity. The experiment was replicated in 2007 by psychologist Jerry Burger, and he got the same results as Milgram.[7] The bullying paradigm has *not* shifted from the 1960s until now. Sociologist Matthew Hollander, in his study of Milgram's experiment, concludes that the "ability to disobey toxic orders" is "a skill that can be taught like any other—all a person needs to learn is what to say and how to say it."[8] The pressing question is: why are we not teaching these skills?

At this juncture, let's remind ourselves about key neuroscientific principles: what the brain does a lot of, the brain gets good at. When cells fire together, they wire together. In other words, we have learned in this experiment that the cells that get fired together in our brains over and over are the cells that tell us if someone has authority—dead giveaway is the white lab coat—then we need to do what they say, regardless of how cruel, sick, twisted, and wrong it is. Our brains, by the time we become random individuals (or average Joes walking around New Haven, where we get asked politely to help out with an experiment being conducted at famed Yale University), have been programed to obey. This is why we have to reconfigure the relationship between our Mind-Brain-Body. We need to *unlearn* helplessness and learn to harness our power, our agency, and our outright refusal to participate any longer in this destructive, outdated, obsolete bullying paradigm.

Not only do we identify with the aggressor and walk around with a Mind-Bully in our skulls, we also bow down to them out in the world, and overriding our own instinctive empathy, we do as they say. But notice: in Milgram's experiment, 35 percent said "no" to the authority figure. It's our choice every single time. We can be in the 65 percent camp who do as they're told, even if it means destroying someone else, or we can be in the 35 percent camp who refuse to participate in an outdated framework that leads to far too much suffering in our world.

Milgram conducted eighteen variations of this experiment, and every single time he got the same results. He concluded in an article, "The Perils of Obedience":

> Stark authority was pitted against the [average Joes'] strongest moral imperatives against hurting others, and, with the [average Joes'] ears ringing with the screams of the victims, authority won more often than not. The extreme willingness of adults to go to almost any lengths on the command of an authority constitutes the chief finding of the study and the fact most urgently demanding explanation.[9]

Can we break the taboo and dare to suggest that the way we raise and educate children, according to the bullying paradigm, leads to compliant adults who do not balk at following orders, even when those orders could result in the suffering or death of someone? Can we take this realization and choose a new paradigm, one informed by brain scans and neuroscientific insights? The brain learns by making mistakes; maybe we can stop punishing them.

Neuroscientist Norman Doidge describes totalitarian regimes like North Korea where children's learning is replaced with indoctrination, little free time with their parents, a cult of adoration for political leaders, and academic exercises designed to instill hate for the enemy, whether it's taught through a storybook or a math problem. What's disturbing is that neuroscientists now see on scans that the brains of individuals who are indoctrinated don't just have set views; they have "plasticity-based anatomical differences" that no longer respond healthily to ordinary confrontation or persuasion.[10] Merzenich discusses the ways in which healthy individuals can transform into "caricatures" of themselves. This occurs through mindless repetition of thoughts, feelings, and behaviors as a person ages, or it can occur through harm to the brain in the cycle of abuse. If an individual is enabled to abuse, like a Harvey Weinstein or a Larry Nassar, so that year after year they repeat the same abusive acts simply with a changing roster of victims, then they turn into "caricatures" of themselves. It's as if they are scripted, and with each repetition, they lose a piece of their dynamic humanity on the outside and their neuroplasticity on the inside. Their "indoctrination," oftentimes through an abusive past, has stripped them of a capacity for true engagement with others. Loving attachment fragments and becomes something akin to "hate for the enemy." As Merzenich writes: "Differences in the quality of 'nurture' regularly make profound differences in physical and functional brains, and in the failures or successes of the people who own them."[11]

Training children to obey adult authority is so influential that it appears to override the brain's natural "mirror neurons"; it dismantles the brain's shared neural circuits that are instantly activated by the pain, emotion, and experience of another person. Mirror neurons were discovered accidentally almost thirty years ago in a neuroscience lab in Parma, Italy. While still a subject of investigation and scientific debate, their existence is substantiated now by "hundreds of studies." The mirror neuron system is responsive even to postures and facial expressions, let alone a body in pain.[12] Neuroimaging research has demonstrated "how powerful the wiring in our brains is for pro-social, or helping behavior."[13] Put that side by side with the average Joes in Milgram's experiment and you have to imagine just how much the bullying paradigm effectively strips us of our innate empathy. One wonders if the training we put children through actually removes or "kills" their mirror neurons so that they grow up to be average Joes who no longer respond with empathy, even to excruciating pain being experienced by others. The natural human impulse to help seems to have been removed through experience or training.

Imagine the wiring that has been laid down in our brains that overrides our mirror neurons and makes us obey even cruel directives. It starts when we are so young and is repeated so often. This is the neural network we must collectively undo. It is going to take an inner revolution (which hopefully becomes a larger social one) to replace obedience to authority with our natural empathy, with our critical thinking, with a deep trust in our instincts, and targeted brain training to reject abusive messaging. As we will discuss in more detail in chapter 6, Merzenich's brain-training programs are an evidence-based remedy for our deficits in social-emotional brain health, among many other areas of the brain that may need healing or strengthening.

If you're suffering from a tendency to obey, it makes it difficult to return to a flexible growth-mindset, so you need to be empathic and mindful with yourself. You can purposefully choose to replace your fixed-mindset with a growth-mindset, as it *is* possible to change your neural networks. Doidge and Merzenich are clear that indoctrination in a bullying paradigm can make your brain rigid and fixed, but they also are international leaders in ensuring you understand that neuroplasticity means you can change. If you're feeling a little demoralized at having to take the hard path of unlearning, let me inspire you with the subtitles of their books. Doidge's books include the subtitles *Stories of Personal Triumph from the Frontiers of Brain Science* and *Remarkable Discoveries and Recoveries from the Frontiers of Neuroplasticity.* Listen to the subtitle of Merzenich's book, which is a documentation of forty years of research into the brain's unprecedented ability to transform: *How the New Science of Brain Plasticity Can Change Your Life.* You too can have remarkable discoveries and recoveries, as well as achieve personal triumph over destructive neural networks that might be holding you back. You too can learn from brain scientists how to strengthen what's weak, make flexible what's rigid, and heal what might be broken.

I asked Merzenich to outline the history of neuroplasticity leading up to his own groundbreaking studies. What follows is the crash course he gave me on the key figures and the way in which they reached science's present-day understanding of the brain's ability to change.

> In the late 19th century the "father of physiological psychology," William James, argued, on the basis of his observational science, that our brains *must* be remodeled by our repetitive experiences. In his view, that remodeling was almost certainly the fundamental basis of the acquisition of all of our neurobehavioral powers, and establishing good "habits" through such remodeling should be the primary goal of human education. Two of the greatest neuroscientists of the first half of the 20th century, the Nobel Laureate Charles Sherrington and the great physiological psychologist Karl Lashley were among a handful of investigators who conducted studies that documented the capacity of the brain of an advanced primate (monkeys or apes) to "remodel" itself following brain injury, or as a result of highly local-

ized, repetitive brain stimulation. In the 1960s and '70s, James' and Lashley's psychologically oriented descendants (L. Roy Johns, Richard Thompson, Charles Woody, Norman Weinberger, among others) documented changes in the neurological representations of electrical or acoustical stimuli used in classical (Pavlovian) conditioning, and recorded physical expansions of the neurological representations of the behavioral responses repetitively induced by that training. Importantly, several of these investigators also showed that these positive plastic changes were reversed when the Pavlovian conditioning (associating a specific stimulus with a reward or a punishment) used in these studies was behaviorally "extinguished." In parallel, other scientists (most prominently, Marian Diamond) showed that enriching the experiences of animals resulted in a thickening of the most directly engaged areas in their cerebral cortex which was accounted for by an elaboration of brain cell processes and an increase in the complexities of their interconnections.

"Case closed," you might have thought. The brain is plastic, for life. Alas, in the mainstream of medical neuroscience, more elaborate and technically more elegant studies seemed to prove exactly the opposite. In innumerable studies of the development of brain wiring, brain connections were repeatedly shown to be alterable by experience or after physical manipulations in the brain in infant animals, but that "plasticity" appeared to come to a screeching halt not long after birth. In a second important class of studies, plasticity processes originally described by the Nobel Laureate Eric Kandel and others were easily recorded in brain slices taken from infant animals, but no such changes could be induced in slices taken from older brains. Plasticity, it seemed, was over and done with in infancy, or in very early childhood. Finally, most compellingly in this scientific epoch, scientists extensively and elegantly studying the maturation of the visual brain (David Hubel and Torsten Wiesel were awarded another Nobel Prize for founding and elaborating these studies) showed that the mapping of the eyes in the brain *was* plastic in the very young life of the brain—but the brain "grew up" (they argued), with all neurons and all brain connection achieving a "mature," hardwired state very early in life. Translated to the human case, in contradistinction to the physiological psychologist's model, plasticity in the medical and educational mainstream was over and done with sometime in the late first or second, or at most the third, year of human life.

My colleagues and I, operating in the medical neuroscience mainstream, conducted important studies that resolved these issues once and for all in the 1980s and 1990s. They (and others) confirmed that two very important things did, indeed, "grow up" in infancy. First, the machinery that *controls* plasticity advances *to take control* of brain change. Plasticity could not be demonstrated to occur in an isolated slice of brain tissue taken from an older brain because in that older tissue, the conditions that govern change were no longer fulfilled at a tissue age for the brain. Change was now only enabled when the brain itself judged that change to be for its benefit. Second, the scale changed with age. While the major trunk-line connections are no longer easily rerouted, the older brain is subject to massive *local* refinement and connectional remodeling.

Importantly, I showed the visual cortical model was exceptional. In all other great brain systems, the machinery of the brain was advanced—specialized—as a result of skill acquisition or improvement, with induced physical and functional brain changes *accounting for* progressions in performance at every training stage. My colleagues and I quickly demonstrated that *every* dimension of neurological representation—place, time, intensity, frequency, stimulus succession—was plastic, at any and at every brain age. We also showed that by following the neurological "rules" being established by our (and by many other concurrent) studies, neurological coding and the machinery that accounts for neurological "representation" could just as easily be degraded as they are advanced by progressive training— and that, within the limits of the rules that govern brain remodeling, brain processes could be changed by training at will. Finally, we conducted studies that helped set up the translation of this science, now deliverable in efficient training forms, to help advance brain health and brain power in struggling and general human populations.

Collectively, these studies provided important scientific underpinnings for translational neuroplasticity, which my colleagues and I carried out of the laboratory into the real world by the establishment of several brain plasticity–based brain health and brain performance training companies. These efforts were recognized by the Kavli Prize, referred to as "the Nobel Prize for Neuroscience," for demonstrating how brain changes all across the life span underlie and enable the growth of our neurobehavioral capabilities.

That decades-long era in science and medicine and education where most experts believed that the brain was "hardwired" long before a child first showed up at school was a kind of dark ages for brain medicine and for educational science. The brain was believed to be a machine that is substantially physically and functionally irreparable, primarily subject to physical regression across your passage in life. To the contrary, the brain plasticity revolution has shown us the brain *gaining* new powers and growing in its health and strength when you use it in appropriate ways. In fact, when you stop firing it up and stop wiring in new skills, the brain *will* falter and wither.[14]

To my mind, this history is the most important for children to learn, far more important than the history of nations, and yet I wonder how many schools teach this curriculum. I was certainly never taught anything about the brain as a child, let alone as an adult in charge of child and young adult learning that occurs in the brain. As an adult, the key takeaway for me from this history of neuroplasticity is to remember that the brain is hungry for learning. Learning turns on genes that change neural structure.[15] This is why no one can tell you what your brain "is." The real story is always about what your brain can *become*. To remind you, every single brain is as unique as a fingerprint. Not even the brains of identical twins have the same nerve branching, connections, and circuitry.[16] The goal is to ensure that your unique brain is always changing, adapting, growing, being challenged, firing up new skills and wiring them in.

Doidge describes the brain's "nerve branching." I find it helps lessen the overwhelming complexity of the brain by remembering that neuroscientists imagine your brain like a tree and use tree-like vocabulary to describe it. Your brain is covered by a thin layer of neural tissue called the cerebral *cortex*, which is the Latin word for the "bark" of a tree. Your neurons have wavy *dendrites* that emerge from them, and this is the Greek word for "branches" or "branch-like." Your neurons or brain cells aren't just covered in branches; they also have *axons* that act and look like the "roots" of a tree. Your *pineal gland*, in the center of your brain, is so-called because it looks like a "pinecone." Your *amygdala*, alarm center for your brain, comes from the Greek for "almond" (seed of a tree) because that's what the neuroscientists think it looks like. Even more amazing, if you open up a walnut, it bears an uncanny resemblance to the two hemispheres of the brain. It's as if nature is trying to remind you that your brain is a growing, tree-like organ. When you breathe in oxygen from trees, imagine the air going into your lungs, which look like a tree turned on its side. One lung is the complex branching root system and the other lung is the dendrites reaching up to the sun. Neuroscientist David Eagleman asks hopefully, "Might we someday be able to read the rough details of someone's life from the microscopic structure etched in their forest of brain cells?"[17]

During two windows of intensive development—the early years zero to three and the adolescent years thirteen to twenty-four—when your brain is capitalizing on all its possibilities, learning, and soaking up information, the neuroscientists describe the phenomenon as "blossoming." Then, when your brain needs to discard excess deadwood—namely, the "excess synapses in the young brain"—and strengthen certain neural networks or branches, they describe this as "pruning." As neuroscientist Sarah-Jayne Blakemore explains: "Synapses that are being used in a particular environment are retained and strengthened; synapses that are not being used are 'pruned' away."[18] One of the best ways to remember your brain's growth potential is to envision it within your skull as a tree that does not stop blossoming and unfurling the leaves of new learning, while discarding or pruning the deadwood of thoughts, feelings, and behaviors that no longer serve your health or happiness. Eagleman puts this ever-changing brain cycle succinctly: "Connections between neurons ceaselessly blossom, die, and reconfigure."[19]

As noted, because your brain has limited cortical real estate, it is quick to oust unused neural networks. It's comparable to Darwin's idea of survival of the fittest happening inside your brain, where synaptic connections are competing to survive and *grow* and this depends on experience and on what you practice.[20] The key word is "practice." Let's pause for a moment to let this concept truly sink in: when you commit to repeated, deliberate, purposeful practice in your life, then you are reinforcing neural networks in your brain. If you practice the piano, you strengthen the neural networks for playing Beethoven sonatas. If you commit time and energy to practicing yoga, you strengthen the neural networks for mindfulness, flexibility, and strength. It's up to you. You do not get good at Beethoven sonatas by practicing the piano for half an hour. Likewise, you do not become adept at mindfulness, flexibility, and strength by going to one yoga class. Practice means repeatedly committing time and energy to some activity. At the beginning, your brain will likely be resistant. It doesn't feel good to prune away all those comfortable, soothing, rewarding neural networks and replace them with awkward new ones that you are not familiar with. However, once your brain has overcome its inertia and resistance, encourage it to let blossom the practices you want in your life and let wither the practices that actually are harming you or holding you back, until your brain prunes them away. What's so amazing about your brain is that once you're on the path to committing consciously, deliberately, mindfully to the healing and healthy habits that will make your life so much better, your brain gets onside. As positive psychologist Shawn Achor explains: "Your brain, once it has tipped toward a habit, will naturally keep rolling in that direction, following the path of perceived least resistance."[21] It gets just as excited and devoted to your new healthy habits as it once was to your old destructive ones. The key is to believe, which triggers the needed "activation energy" that catalyzes the deliberate practice of your new way of being.[22]

Whatever you give your time, focus, and energy to, your brain responds to it by laying down neural networks to support it in more and more sophisticated ways. Turn this around and understand that whatever you *don't* give time and energy to, your brain responds to it by "pruning" those neural networks away.[23] As Merzenich clarifies: repeatedly firing up a neural network works in concert with pruning away a neural network. "Positive plastic brain changes work to create a brighter and sharper picture of what's happening. At the same time, negative plastic brain changes are erasing a little of that irrelevant and interfering haze or noise that frustrate the construction and recording of a clear picture."[24] The healthy brain depends on clear pictures and suffers when it is perpetually distracted and pulled in different directions by "haze" and "noise."

When you have a fixed-mindset, then you see your brain as if it's a machine. When you have a growth-mindset, then you understand your brain as if it's a growing tree. Psychologist Carol Dweck studies fixed-mindsets and growth-mindsets. The former holds you back, and the latter allows you to optimize your brain's performance. To encourage growth-mindset, it is important to drop labels altogether. According to Dweck's research, the complimentary labels are just as misleading and inaccurate as the put-downs. When children with a fixed-mindset—whether believing they are smart or slow—are presented with an obstacle or challenge, they give quickly. Children with a fixed-mindset try to avoid having to face obstacles and challenges at all. The ones labeled "intelligent" are drawn to activities that they can master to confirm their gifted label. Kids labeled as "struggling" or "slow" fear activities that are challenging because they feel ashamed at their inability to quickly master them. Kids with fixed-mindsets do not associate overcoming obstacles or rising to challenges with *practice*, effort, and work. In contrast, kids who learn according to a growth-mindset are drawn to problem solving; they roll up their sleeves and get to work. If they struggle, they simply work harder; they get creative; they try different strategies; they think outside the box. Regardless of where they are in their learning journey, these students believe that practice, effort, and work lead to success, so when they come up against a wall of some kind, they work double-time to try and find a way to scale it.

Psychologist Angela Duckworth's work reveals that having a "fixed mindset about ability leads to pessimistic explanations of adversity, and that, in turn, leads to both giving up on challenges and avoiding them in the first place." The key is to maintain, even in the face of adversity, a growth-mindset that allows you to see suffering as a kind of crucible where hard lessons are learned and your ability therefore improves. This is where we can turn post-traumatic stress disorder into post-traumatic growth. If we never forget that our brains have unprecedented neuroplasticity, we can hone the superpower of "grit." As Duckworth has shown, grit leads to perseverance, and that means you will seek "out new challenges that will ultimately make you even stronger."[25]

In her research, Carol Dweck argues that we need a complete reversal from blaming, criticizing, humiliating for mistakes, complaining about effort, insisting on one way to do things. Although we're told it's for our own good, Dweck's research has shown that this kind of bullying approach undermines learning each step of the way. She explains that if adults "want to give their children a gift, the best thing they can do is to teach their children to love challenges, be intrigued by mistakes, enjoy effort, seek new strategies, and keep on learning."[26] "Keep on learning" could be your mantra if you want to keep your brain healthy and if you want to change from a fixed-mindset to a growth-mindset. If you want to maintain a healthy brain into old age, being a lifelong learner is the game-changer.

Instead of beating yourself up when you make mistakes, why don't you try being "intrigued" by them? If you have tried to stop destructive behaviors, failed, and feel like giving up, dust yourself off and "seek new strategies."

"Keep on learning" is the opposite of the fixed-mindset that leads to radicalization at worst, defensiveness and inflexible brains at best. Dweck asserts that "school cultures often promote, or at least accept, the fixed mindset. They accept that some kids feel superior to others and feel entitled to pick on them. They also consider some kids to be misfits whom they can do little to help."[27] This is an apt description of the bullying paradigm that hinges on fixed-mindset and likes to assign labels without acknowledging that the brain is as organic and able to grow as a tree. Dweck shares that "some schools have created a dramatic reduction in bullying by fighting the atmosphere of judgment and creating one of collaboration and self-improvement."[28] Rigid brains and inflexible mindsets—whereby individuals and whole organizations ignore what is different and even outright reject it if it doesn't fit their established pattern of thinking—cause much of the world's suffering. It is a neuroscientific way of describing racism or sexism. Entrenched belief systems, resulting from rigid brains, put independent, different, dynamic brains at risk. As noted, until the day we die, we can change our brains; however, the more rigid and fixed our brains become, the harder it is to change them. The more rigid and fixed our brains become, the likelier it is that we transform into caricatures of our once healthy selves.

Doidge explains that as we age, the plasticity or malleability in our brains declines: "It becomes increasingly difficult for us to change in response to the world, even if we want to." An individual who has been bullying or abusing for years, when confronted, may no longer have the brain flexibility to see they are being hurtful and respond by changing their ways. It's not impossible, but each passing year that reinforces the neural network of destructive behaviors will make it more and more difficult to change, likewise with habits or behaviors that we engage in that hurt us. Research "shows we tend to ignore or forget, or attempt to discredit, information that does not match our beliefs, perception of the world, because it is very distressing and difficult to think and perceive in unfamiliar ways."[29] Some people prefer alternative facts to hard facts that can be objectively measured and verified because their brains have become rigid; their mindsets have become fixed. They struggle to think and perceive in unfamiliar ways. They prefer to obey and do what they're told by a chosen authority.

You may have struggled with the last action step that asked you to be a writer, rather than a reader, of culture. Do you see that choosing new beliefs and behaviors is really hard for our brains? Maybe the reason we find it so hard to exit the bullying paradigm that many of us grew up in is exactly because it has shaped our "perception of the world," and now our brains are resisting and struggling because we're asking them to "think and perceive in unfamiliar ways."

You need to take your brain by the hand and introduce it to this new way of being. You're going to have to work hard to make the unfamiliar familiar through daily practice. It's not a one-shot deal. No quick fixes. What may happen is that your brain starts to feel a lot of distress, and this may make you fall back on familiar ways of self-medicating your stress, anxiety, and pain. And then you falsely believe you can't break the cycle of destructive behaviors. It's maddening when you think about it. Just remember right here and right now: you have to believe in your Mind-Brain-Body because once they get in sync and work together, then nothing can stop you. Overcoming your brain's distress requires actually acknowledging it and then calming it down, not once but over and over again. Once your brain is calm and feels safe, you can begin the hard but rewarding work of brain training.

Helen Reiss cautions against seeking external rewards. Your brain reacts to external rewards by experiencing a hit of the neurotransmitter dopamine, which in simple terms gives us a jolt of feeling "happy." In a bigger sense, these extrinsic rewards communicate to your brain that you are part of a social circle and your brain knows this is important for survival: "The flood of neurochemicals creates cravings for these bursts of attention and makes us constantly attend to our phones to experience the next 'hit.'"[30] A one-year-old does not seek external rewards. He does not look to others when he is learning to walk; he is laser-focused on his own growth-mindset that leads him to try every trick in the book to make the leap from crawling to walking on his own two feet. Merzenich stresses how natural this learning process is for the brain: "Standing enables cruising. Cruising enables walking. Walking enables running. Running enables jumping over hurdles or through hoops."[31] As you learn a skill or hone an ability, your required brain processes become more and more precise and reliable. We all want to be able to jump over the hurdles and through the hoops thrown at us throughout life. It is a process, and you need to trust yourself to put in the required effort as this is exactly what your brain supports. "As the actions of neurons in the brain become more coordinated, the machinery of the brain detects that growing teamwork and produces chemicals that enable performance achievement at the next highest levels."[32] This is an apt description of what's happening inside the skull of growth-mindset, which manifests as someone working hard at meeting a challenge, mastering a new skill, or pushing through obstacles to reach their goal. Now imagine how this innate capacity in the brain to achieve mastery gets frustrated by the noise of abuse. Imagine the interference in the brain's natural way of learning if someone is constantly stopping you to yell at you, put you down, publicly humiliate you, berate you, and threaten you. These are exactly what's in the educational toolbox of a Terence Fletcher from *Whiplash*. They are normalized approaches in the bullying paradigm. The goal is to walk away from external rewards or external abuse that dominate a fixed-mindset.

Dweck advises to strive for intrinsic rewards from your Mind-Brain-Body. As soon as you seek praise from others, you could be setting yourself up for failure. Dweck provides an example of a parent who speaks to his child after she failed to earn a ribbon in her gymnastics competition with the goal of igniting her intrinsic reward system.

> Elizabeth, I know how you feel. It's so disappointing to have your hopes up and perform your best but not to win. But you know, you haven't really earned it yet. There were many girls there who've been in gymnastics longer than you and who've worked a lot harder than you. If this is something you really want, then it's something you'll really have to work for.

This father also let his daughter know that if she wanted to do gymnastics just for the fun of it, that was okay too, but if she wanted to excel at competitions, she'd have to put in the work.[33] Dweck concludes by noting that Elizabeth was very competitive and actually wanted to win. So she put in a great deal of time to improve her skills and especially focused on her weaknesses. In the next international meet, she won five ribbons and took home the trophy as overall champion of the competition. All of us can achieve our best with *intrinsic* rewards and a growth-mindset that knows working hard is how we reach our goals.

To heal your scars, restore your health, and fulfill your potential, it is vital to debunk the myth, generated by the bullying paradigm, that harshness or unempathic conduct builds toughness, grit, and resilience. Scientific research provides extensive evidence that the opposite is true. This is why you won't find any believers in the bullying paradigm speaking about science and especially not neuroscience. When children are on the receiving end of consistently nurturing, responsive, respectful treatment, they develop security and a deep-seated resilience that supports them in identifying and regulating their own challenges.[34] If you did not grow up in this kind of environment at home or at school, you can re-create this kind of healing and restorative learning environment yourself. If in the workplace you've been exposed to a toxic environment that has eroded your neural networks and your health, you can re-create a healthier, more restorative environment by exiting the bullying paradigm.

While the bullying paradigm creates destabilizing, fear-inducing, unpredictable conditions, the cornerstone of learning and healthy brain wiring is safety. So you're going to need to learn over and over again that creating a safe environment that is consistent and calm is the foundational step to laying down new, stronger, desired neural networks. Activating your growth-mindset adds to this safe environment because it establishes that learning by making mistakes—and repeating mistakes—is how your brain grows and discovers its considerable abilities, power, and inventiveness. You may find that you need a teacher, mentor, or coach. This kind of authority figure is the opposite of the one that Milgram studied in his experiments at Yale University. His authority figure issued commands that pushed individuals to shut off their curiosity, their critical thinking, their ethical and empathic centers. Milgram's authority figure controlled and commanded the average Joes to do as he said, regardless of the suffering and danger being caused to the victims.

In this action step, you will seek a different kind of authority figure to replace the Mind-Bully who demands compliance. Instead, you will seek a teacher, coach, or mentor who listens to you, encourages you, and ultimately empowers you. Like the Mind-Bully, this authority figure can be in your own skull. In the action step for this chapter, we discuss this ideal figure to help you develop your growth-mindset and thereby grow your talent.

# Step 3: Grow Your Talent

Journalist Daniel Coyle discovered in his international exploration of "talent hot-beds" that there are three key ingredients: ignition (the belief that you can achieve your goal); deliberate practice (the commitment to working daily at improving your skills that assure you can achieve your goal); and finally, being coached by a "talent whisperer."[35] Coyle used British runner Roger Bannister's triumphant achievement of breaking the seemingly impossible four-minute mile in 1954, which was repeated within weeks by others. Coyle's point is that if you set yourself a goal that doesn't seem realistic or even possible, you'll struggle to achieve it. To grow talent, it's important to start out with a believable goal. To light a fire within you, necessary for the hard work and commitment needed for deliberate practice, belief in yourself is key. One of the most destructive legacies of the bullying paradigm is that it removes your belief in yourself. It replaces belief in yourself with inaccurate labels, humiliating put-downs, and a sense that you deserve abuse and that it's your fault and that you're the one responsible. The bullying and even abusive individual frequently walks away with few consequences, sowing doubt in your mind about what happened and who is to be held accountable.

The most important part of this action step is to identify when and how you might have lost belief in yourself. Then you need to strategize and develop a plan for restoring it. You have neuroplasticity so you can change your brain to fulfil a different path for yourself. Only when you truly believe in yourself can you find the energy and commitment to dedicate time and effort to practicing.

In his study of deliberate practice, the second phase of talent growth, Coyle draws on neuroscientific findings that show when we practice an activity, it insulates our corresponding neural networks with myelin. We've mentioned myelin before. To remind you, it's the fatty insulator that wraps around an axon or neural wire when that wire is fired up or activated over and over again. As you practice something, and thereby acquire knowledge and skill, within your skull, the matching neural networks are becoming more and more efficient and faster and faster at sending electrical signals due to the myelin. As Merzenich puts it: "Wires with better insulation ship out more-coordinated information at higher speed."[36] Do a quick overview of what axons in your brain are insulated with myelin that makes them highly efficient and quick. If you practice compassion daily, the neural networks for this behavior are myelinated in your brain. In contrast, if you bully people daily, then the neural networks for this kind of conduct are myelinated in your brain. You might have myelinated the axons needed to read and absorb information, speak multiple languages, do chemistry experiments, play the tuba or chess, throw a discus, shine at video games, or negotiate or navigate or whatever. Quick reminder from chapter 1: what fires together wires together; what the brain does a lot of, the brain gets good at. In Coyle's study of this neuroscientific finding—that myelin is integral to how we "grow" talent in our brains—he offers specific details on how to practice, as it can't be simply repeating a particular move and expecting to improve. This is where excellent parenting, coaching, teaching, and mentoring come in.

Coyle says to truly fire up talent in the brain at the highest levels, it's a game-changer to have an exceptional coach who, as noted, is a "talent-whisperer." The qualities of a talent-whisperer are as follows, and you'll note (especially with my commentary in parentheses) that these qualities are the opposite of those that rule in the outdated, bullying paradigm:

- "quiet, even reserved" (in contrast to the yelling, swearing, berating, humiliating, threatening that the bullying paradigm keeps arguing is key to success);
- a "steady, deep, unblinking" gaze (in contrast to those who bully who struggle to meet your eyes as documented in research);
- "listened far more than they talked" (in contrast to the bullying paradigm that gives lots of airtime to those who taunt, put down, shame, and rarely stop with the verbal onslaught); and
- "an extraordinary sensitivity to the person they were teaching, customizing each message to each student's personality" (in contrast to bullying behavior, which is defined as insensitive, unempathic, dependent on a power imbalance that targets multiple victims whose personalities are irrelevant and simply serve the repeated pattern of abuse).

The final quality is exactly what constitutes an Empathic-Coach. This figure is responsive, not reactive. In contrast, the Mind-Bully is highly reactive. To remind you, the Mind-Bully is reactive because it likely developed as a defense strategy in the bullying paradigm. The Mind-Bully can lead you to feel anxiety and depression. It can push you to soothe or self-medicate chronically stressed feelings with substances or self-harm. These are internalized reactions to bullying and abuse. The Mind-Bully can also lead you to be aggressive, raging, lashing out, and humiliating or harming others. These reactive behaviors occur when you externalize a history of being bullied or abused. Both of these patterns of behavior might make you feel helpless to change. You might feel trapped and as though you can never escape. But neuroplasticity, which underpins a growth-mindset, begs to differ. There's a vast amount of evidence that shows you can change your neural networks and thereby change your reactive nature into a much more mindful, responsive one. It's not easy, but it *can* be done. First, you need to oust the Mind-Bully and make way for the Empathic-Coach. You need to free up cortical real estate for the Empathic-Coach to take root in your brain.

Why is it worth all this trouble to make the switch from what you may well have experienced and know (the Mind-Bully) to the unknown (Empathic-Coach)? Because the Empathic-Coach is a talent-whisperer. These coaches, with their quiet observation, their ability to listen with empathy, their skill at tailoring their message to you and what you need in the moment, are golden. It's not easy to make the shift from the established neural network in your brain that may default to the Mind-Bully and begin charting new territory, trailblazing new paths, and redrawing outdated maps with new ones, but it's worth it.

Daniel Coyle notes that talent-whisperers, like one of the greatest coaches of all time, legendary basketball coach John Wooden, spend next to no time on compliments or criticism. Remember that Carol Dweck confirms that complimenting a learner's quick wit or critiquing a learner's slow answer are both counterproductive. That's why talent-whisperers offer responsive information without passing judgment.[37] When you choose to internalize this figure, rather than the Mind-Bully, then you need to constantly remind yourself to be kind and curious (namely empathic and responsive). Don't judge. Just listen and observe your Mind-Brain-Body as they work together. They will reward you with tapping into reserves of talent you might not have known were even there. Remember, there are no quick fixes, and that's why Wooden's mantra is "The deeper you practice, the better you get."[38] The bullying paradigm will be quick to tell you that Wooden probably lost a lot of games and championships because he failed to abuse his players, but like everything else you hear in the bullying paradigm, there's no evidence to back this up. It's false. Coaching the UCLA Bruins over a twelve-year period, Wooden's teams won ten national NCAA championships.

## CHAPTER 4

# Exit the Cage of Learned Helplessness

## *Step 4: Unlearn and Rewire*

Psychologist Martin Seligman conducted an experiment using dogs. German shepherds were caged and given repeated electric shocks that caused so much pain the animal howled and desperately tried to escape. But after days of suffering, the dogs gave up, and even when put in a cage without restraint and an obvious escape route, they remained helpless, unable to leave. John Medina comments that "most learning had been shut down, and that's probably the worst part of all."[1] Seligman's experiments on German shepherds, horrendous as they were, resulted in the important discovery of learned helplessness. You don't need to be the target of bullying or abuse to develop learned helplessness. Simply witnessing abuse and bullying is enough to make you feel humiliated, anxious, and trapped. These kinds of toxic feelings are not healthy for your brain, especially when they occur repeatedly, becoming chronic. Neuroscientist Rick Hanson writes that just "a handful of painful experiences of futility can rapidly become a sense of helplessness" that is linked to "depression." While that's depressing news unto itself, the good news is that we can rewrite those moments of futility into powerful moments of efficacy, turning our depression into a sense of competence and confidence.[2] Living in a home, attending school or work, playing sports, or performing arts under bullying or abusive conditions leads to deep feelings of "helplessness and powerlessness."[3] As psychologist Angela Duckworth states: "The scientific research is very clear that experiencing trauma without control can be debilitating."[4]

In 2008, when Montgomery's younger brother, Angus, was in third grade, his teacher called us to come into the school for a discussion; she had some concerns about Angus's learning. While he could read out loud perfectly and was animated and engaged in class discussion, his test scores were hovering around zero if he had to answer questions about the story they had read. Instead of demonstrating his knowledge of the book, Angus would simply invent completely different material as if he had never even read the story. I found this surprising because he was fascinated by Greek and Roman mythology and could chatter on in great detail about Zeus and Athena or tell you about the monsters in the *Odyssey* and the battle scenes in the *Iliad*. I had read these stories to him, along with many others, and he never forgot them. They were part of his ever-present, working memory. So what was this about not being able to produce answers on a test in grade three on a child's book?! The penny dropped. This sounded an awful lot like what my brother had suffered with in school. Except now, we could take Angus to an educational psychologist and get his information processing assessed. After the assessment, Angus presented as having less than 1 percent brain capacity to retain visual information, while he was off-the-charts exceptional in retaining auditory information. The problem was he was *looking at* the text of the book in grade three, not *hearing it* being read to him.

Much of the present-day school system and workplace revolve around visual processing, and therefore this assessment was concerning to say the least. But what could have been a significant blow in educational terms was transformed by Eaton Arrowsmith's program that was constructed on neuroplasticity and offered cognitive rewiring. "Cognitive" is a catch-all term for reasoning, thinking, and remembering. It's about intellectual processing in your brain. Now, imagine your brain as being full of neural networks that look like stars connected by wires. If we looked through a telescope at the night sky of our son's eight-year-old brain, we'd see that the constellation or wire for visual processing was not in good shape. It had a bunch of "breaks" in it that blocked the flow of information.

Despite being a teacher myself, I had not even known there was a program available that strengthened deficits in the brain like "less than 1 percent capacity to process visual information." A learning resources expert told me to check out Eaton Arrowsmith School. I had attended school and then attended university for eleven years, completing a double BA, an MA, and then a PhD without hearing anything about my brain or anyone else's. I was in the business of brain activities like interpretation, analysis, critical thinking, writing, articulating original work, teaching, and learning, but still managed *not* to know anything about brains, my own included. If you think about it, it's pretty funny that in the education system, we rarely, if ever, mention the brain, let alone learn about it. Teachers trained by the faculty of education learn about teaching, do practicums, and earn certification without so much as hearing the fundamentals of the brain, let alone being informed about the organic learning-center they will be in the business of working with. It would be like doctors specializing to become cardiologists and not actually studying and learning about the heart.

With a lot of worry and some doubts, in the following year, we sent Angus to Eaton Arrowsmith School, an applied neuroscience program that next to no one had ever heard of. They offered an innovative curriculum focused on "strengthening cognitive deficits." We were worried because it provided a completely unheard-of curriculum, which was almost too complex to understand, plus it was very expensive. We had never heard about a school that used targeted brain training, yet in neuroscientific circles, this approach is seen as obvious and straightforward. Neuroscientist Sarah-Jayne Blakemore poses a rhetorical question: "Is it possible to train the brain?" and then answers brightly: "Of course it is! The brain is plastic, capable of change, and the brain changes whenever you learn something new."[5] It is this gap between life as we know it and what has been discovered in neuroscientific labs around the globe that allows the bullying paradigm to flourish. Not knowing about the brain, and what changes it, means being ignorant about how all forms of bullying and abuse negatively change and thereby harm brains.

Did we see results with the seemingly eccentric Eaton Arrowsmith approach? After about three months, Angus got into my mom's minivan and said, "Why did you ever have me at that other school where I couldn't learn anything?" The other school he had been at was excellent, but it was *not* able to teach a brain like his that was "disabled" or exceptional in how it processed visual information. The Eaton Arrowsmith program specializes in rewiring students' brains to repair cognitive deficits, but it costs thirty thousand dollars a year. What if we couldn't afford it? What if it didn't work while we poured our retirement savings into it?

Throwing caution to the wind, we had taken a significant risk, and we watched firsthand the amazing, positive force of a new neuroparadigm at work. After four years, Angus was again assessed by the same educational psychologist who did the initial test on him, and she was shocked to see that he no longer had a visual retention deficit. In fact, on top of being a gifted auditory learner, he now has the capacity to read text on the page with his eyes and retain detail at an extremely high level. Today, he reads voraciously and speaks knowledgeably about hundreds of books. As documented and outlined by Steve Silberman's *NeuroTribes*, in the future, our son's experience won't be prohibitively expensive for other families, nor will it be seen as unusual. Silberman makes a compelling argument for, and thereby imagines, a "future of neurodiversity."[6] My big takeaway from this experience was that these constellations or wires in our brains can be changed or "rewired." Who knew? I was suitably impressed with the eighty-six billion neurons that glittered like stars in the skull but was a little overwhelmed by the idea that the constellations could be moved, shaped, and changed negatively and positively.

As a mother, I was anxiously hopeful that Eaton Arrowsmith could fulfill its promise to rewire our son's visual deficit, but as a teacher, I was in free fall. It changes *everything* to think about students' brains as being shaped by you (along with many other factors). As a teacher, you're not pouring knowledge from a great fountain into your students' brains. You're more like a gardener pulling up weeds, nurturing the soil, protecting saplings and blossoms from heavy rains and high winds. You're pruning away the dead branches and strengthening the strong ones that will help your students flourish at whatever it is they want to do in life. In *The Gardener and the Carpenter*, psychologist Alison Gopnik writes about children and their brain diversity and our relationship with them. She advises parents to create environments for their children to grow in creative, plant-like, unpredictable ways and not to hem them in with buildings or constructs that require them to fit into established, rigid rooms.[7] This echoes the growth-mindset researched by Carol Dweck and reminds us to think of our neural networks in the way that neuroscientists do, namely as comparable to a vast forest of trees.

What would have happened to Angus if, like my brother in the 1970s, he was made to feel that his inability to remember visual information was a sign that he wasn't smart? What if his teacher had told him that he was lazy or unfocused when he did poorly on tests, when he made up answers instead of recorded what was in the book? What if we as parents had believed he was "slow" or lacked intelligence or "wasn't trying"? These kinds of inaccurate labels trap individuals into believing they are fixed, stuck, and helpless. Many adults still believe these assessments or judgments affixed to them during childhood. You have the opportunity to take the discovery of neuroplasticity and get inspired about optimizing the potential of your brain and strengthening it through brain training to fulfill your potential. At any time, you can exit the cage of learned helplessness.

Cofounder of the school Barbara Arrowsmith-Young herself was challenged by a "learning disability" and in response developed a whole new way of thinking about brains and education. In her memoir and educationally groundbreaking book, *The Woman Who Changed Her Brain: Unlocking the Extraordinary Potential of the Human Mind*, Arrowsmith-Young shares case histories of the children and adults in her program who, like her, have changed their brains and thus their lives.[8] Perhaps you or others you know have extraordinary potential tucked away in *your* brain that may have been hidden due to the bullying paradigm in which so many of us have had to operate since childhood. How was Arrowsmith-Young bullied? She was called "stubborn," "slow," and "retarded." As noted, it was not known at the time that these kinds of labels negatively impact brains, but it is known now. Ironically, it was the bullying approach of the school system that was in fact slow and backward. Likewise, it is in general the school system (of course there are amazing exceptions) that stubbornly clings to an outdated paradigm when it comes to education and learning. Arrowsmith-Young was a gifted inventor, an extraordinary thinker, who has dramatically transformed the brains of others for the better. The school system passed judgment on her, applied labels to her, and she rejected them, trailblazing her own path that led her to discover neuroplasticity at a time when it was emerging as a new understanding. Norman Doidge tells the story of Arrowsmith-Young. He explains how in this program, students can gain "access to skills whose development was formerly blocked" and, in reaction, not only feel belonging and safety but also "feel enormously liberated."[9] Neuroplasticity means all of us can exit the cage of learned helplessness created by the bullying paradigm.

I learned about the potential of neuroplasticity by watching it happen before my very eyes, in a high-stakes educational situation with my own son. I knew from my experience as a sister, and then multiple times as a teacher, how remarkably hard, frustrating, and demoralizing it can be to have to learn in an educational world that does not match the way your brain works. Our son was spared this misery. And what a discovery. The brain, with focused work and repeated practice at timed intervals, can radically *change*. The implications of neuroplasticity, or your brain's ability to change, are exciting and far-reaching. In short, it means you have the capacity to heal neurological scars and restore your holistic health that aligns your Mind-Brain-Body. Neuroplasticity also meant that while Angus was strengthening cognitive deficits, his older brother, mouth scorched by cortisol, was suffering a profound weakening and scarring of his cognitive capacities due to frequently repeated abuse.

Doidge talks about a boy who, like our younger son, was lucky: "Once he was liberated from his difficulties by Arrowsmith's exercises, his innate love of learning emerged full force." He speaks about children who suffer the kinds of brain challenges Arrowsmith-Young had to contend with herself—despite her brilliance—and he comments: "These children are often accused of being careless." "Such children often forget instructions and are thought to be irresponsible or lazy." "They often appear disorganized, flighty, and unable to learn from their mistakes." In contrast, "Barbara believes that many people labeled 'hysterical' or 'antisocial' have weaknesses" in the frontal lobe, which is the site in the brain for planning, developing strategies, sorting out what is relevant, and forming goals and sticking to them. Notice the liberating shift from a moral assessment to a neuroscientific one.[10]

Children who are struggling to learn just might *not* have deplorable characters that earn them bullying labels like "laziness," "hysteria," "flightiness," or "irresponsibility." It would be obvious to a neuroscientist that there is some sort of deficit or injury within their brains. Maybe it's time for us to change up the labels that offer moral assessments and replace them with labels that offer brain insights. If someone has a broken arm and can't write, we do not say she's lazy.

In contrast, if someone has a "broken" brain and can't write, someone might still say that individual is *not* trying, needs to work harder, is sloppy, has terrible penmanship. Someone might even say the person is lazy. What's the difference between these two scenarios? One is visible. We can see the broken arm. The other is invisible without a brain scan.

Merzenich's "computerized, brain plasticity-based, game-like training programs" have set several million children with impaired language and reading onto the path of academic success at minimal cost.[11] Unfortunately, far too many children who struggle to read "associate language with trouble" due to the "ridicule and negative feedback" they receive from the adults in their world, whether parents or educators.[12] Merzenich recounts the brutally sad story of John Corcoran, who was paddled at school because he couldn't read. First, the teacher, clearly a proponent of the bullying paradigm, would hit him, and then each student was encouraged to strike the back of his legs before he was sent back to the "Dumb Row." This is a prime example of the ways in which adults "teach" children the ways of bullying, but when children mimic their cruel and humiliating targeted abuse of a child, we turn around and say, "Don't bully," and express concern that bullying is reaching epidemic proportions in child populations.

When my brother and I were attending school, teachers didn't know that a student like John had dyslexia so that reading presented specific challenges for his exceptional brain. Ultimately, John learned to read in midlife and wrote a book about it.[13] His story illustrates Merzenich's crucial message that our brains are plastic and can learn—and *want* to learn—until our final moments on the planet.[14]

More than ten years ago, Doidge said that "the Arrowsmith approach, and the use of brain exercises generally, has major implications for education."[15] How many schools use brain exercises today? Next to none. It is vital that doctors who want their patients to be healthy are knowledgeable about the body. It is equally vital that parents, teachers, and coaches who want children to fulfill their learning potential are knowledgeable about the brain. A quick glance at the workplace reveals a similar disconnect whereby employers have fitness rooms to keep bodies healthy but do not put the same investment into employee brain health. If cognitive deficits in the brain can be rewired, so can disorganization, antisocial behavior, hysterical conduct, and other brain "breaks."

French educator Paul Madaule has set up thirty centers across the world that work successfully with children who have brain challenges. Doidge tells the story behind why Madaule devotes his life to healing students' brains. In 1960s France, just before my brother and I were in school, next to nothing was known about dyslexia and other brain disabilities or exceptionalities that Madaule himself suffered, like Corcoran and Arrowsmith-Young did. And as I witnessed, the school system certainly didn't know anything about this kind of "neurodiversity" in the 1970s either. Madaule would get a report card weekly that positioned him at the bottom of the class and failing every subject. His parents thought he was "lazy," and there would be weekly "screaming matches" when they had to sign his reports. And the children "teased" him. His "physical education teacher picked up on the mockery and called him *une oie grasse*—a fat goose."[16] Perhaps we can forgive the school system for the barbaric, ignorant report cards. Perhaps we can relate to and understand parents who are misguided, worried, and angry. But imagine a teacher who sees a child targeted by peers, clearly struggling, and chooses to mock him. That is notably cruel adult bullying. That is emotional abuse. It hurts brains. Imagine how impossible it would be for Paul's brain to grow, heal, and learn when his brain was on high alert from anticipating daily humiliation and suffering.

The most difficult obstacle you may face in escaping the cage of learned helplessness is that unlearning is so much harder than learning. In childhood, your remarkable brain, primed to grow and develop, may well have soaked up all the directives of the bullying paradigm. It has become so normal that it's difficult to unwire, unlearn, and escape. The bullying paradigm has transformed far too many into individuals who struggle to question, let alone resist the most blatant abuse. Just like victims, those who abuse are trapped in a cycle. Paul Pelletier explains: "Many of today's leaders were mentored themselves by command-and-control managers, and the culture of a lot of organizations is still based on command-and-control norms. It is hard to escape this leadership style's historic influence and dominance."[17] Another way to say "hard to escape" is with Seligman's psychological term "learned helplessness." It's one thing when an individual suffers from it. It's quite another when our society as a whole suffers from it.

Most of us would agree that a child or even adult who is being abused or bullied may feel that there is no escape. That victim may well feel utterly helpless. And yet the first question a lawyer will ask a victim is: "If you were being bullied or abused, why didn't you leave?" Here's the problem: few, if any, children or even adults know how to respond to that question with brain-informed terminology along the lines of explaining that they had a "perception of inescapability" and its "associated cognitive collapse."[18] The lawyer may then say: "The fault lies with you because you are an addict." Again, the victims are unlikely to have the research before them that allows them to explain their substance abuse is in fact an "attempt to escape distress," that they are "self-medicating conditions like depression, anxiety, post-traumatic stress or ADHD."[19] Frequently, victims of abuse lack the education and training to properly identify and convey their condition. They suffer from "learned helplessness" or the belief, perception, thought that there is no escape. What accompanies this shutdown in the brain is the collapse of thinking or cognitive skills. If you can't even think straight or problem solve, how are you supposed to design an escape plan that gets you away from your tormentors? This becomes particularly dangerous when you internalize the aggressor so that the tormentor, or Mind-Bully, takes up space in your head.

Neuroscientist David Eagleman is clear that for the brain, "proper development requires proper input." If the input is abusive or neglectful, the brain does not develop properly. He refers to Harry Harlow's research, which is as heartbreaking as Seligman's research on German shepherds. Harlow used monkeys to try and figure out depression. While he didn't use active pain like an electric shock, he did something just as damaging: he put the monkeys when they were babies into a cage that was empty and therefore had no mother. Eagleman writes, "Because the baby monkeys never had the chance to develop normal bonds (they were put in the cage shortly after birth), they emerged with deep-seated disturbances." These disturbances shaped their adult lives so that, as we've seen in the bullying cycle, they became withdrawn, neglectful parents or violent, aggressive parents. "The results were disastrous."[20]

This is a moment to close your eyes, take a deep breath, and allow yourself to feel some empathy, some heartfelt compassion, for anyone who feels trapped in the cage of bullying and abuse. That empathy, that compassion might be directed at yourself, and if so, that's step 1 to stop the cycle of beating yourself up for getting stuck; for not finding a way out sooner; for feeling paralyzed, which can go on for years. All of those feelings are to be expected when you've been bullied or abused. They are natural brain responses to abuse. So no more blaming yourself. Let's put your energy and focus into walking away, head held high, from the cage of learned helplessness that simply cannot hold you.

What's hopeful and exciting for you and for all who have been mislabeled—and might even have internalized these false statements about your own Mind-Brain-Body—is that neuroplasticity provides us the ability to erase negative labels because we in fact have an immense capacity for rewiring, for unlearning, for building new neural networks. We can strengthen our brains. In her 2014 TEDx Talk on "Potential," Britt Andreatta says that we need about forty repetitions before we lay down a new neural pathway. Andreatta is a sought-after performance coach, and in her TEDx, she shares a story about a client who was alienating his colleagues at work with his desperate need to be seen as smart. Turns out that in childhood, this man's need to belong got derailed due to bullying. Andreatta recognized his frantic desire at work to be heard and to be valued as coming from a past "break" in his brain. As a new student, he was inadvertently exposed by his teacher, who threw him right into a spelling bee as he entered the class. He got the word wrong and became the laughingstock of the other children. They bullied him as "dumb" for years. He was pushed into the out-group by the children, which is extremely distressing since wired into our brains from the start, as our survival hinges on it, is the human drive to belong. The neurological scars from bullying impacted the man's professional life in negative ways for years. Andreatta concludes her TEDx Talk with a call to arms: "Let's teach this stuff!" And by "stuff," she means neuroscience.[21]

When you resist learning, when you don't want to gain knowledge, when you quit trying, you need to hold back the labels and instead think like a neuroscientist. Resist the temptation to put a moral label on how you're feeling and your sense of failure. Instead, you need to assess your struggle from a brain perspective. Just because you cannot see your brain with the naked eye is no reason to assume that it is operating at full capacity and is able to optimize its performance. My goal is for you to not put yourself down and not to live your brain is static, a done-deal, immovable, because that's not true. These are the bars that create the cage of learned helplessness. Liberate yourself. You can rewire your brain by what you practice deliberately. And you can put into the illuminating light of science certain moral labels that were assigned to you in childhood, during your teen years, or even during adulthood that might be suitable for the bullying paradigm of the twentieth century but are most definitely out of date in the twenty-first-century *neuroparadigm*.

Merzenich exclaims that not only can we train our brains to be more skilled at sports or chemistry or a second language; we can also transform ourselves ethically and spiritually right until the end of our lives. "We all have it within us to rise to a higher plane. We have the capacity, as long as we're alive, to change the higher operational principles that guide our very soul—to change the person that we are, for the better."[22] He explains that children, like our son Montgomery, "who have a brain that has been disadvantaged" by an abuse history can still recover and succeed. He explains that children, like our son Angus, who have a brain that has been disadvantaged "by genetic weakness aren't stuck on the inalterable path to failure."[23] He also questions the very inevitability and inescapability of the bullying paradigm. If we can enhance our conduct and rise to a higher plane, then perpetrators of bullying and abuse, especially if identified early, can be healed and their brains can be restored to health.

We found out in 2016 that Angus suffers from a rare genetic condition called Klippel-Feil Syndrome (KFS), which resulted in the fusion of his two top vertebrae and led to cognitive deficits. By the time the specialists reached a diagnosis, he had already undergone the brain training with Eaton Arrowsmith and repaired the "breaks" in his brain. How many parents don't even know this is an option? Merzenich is very clear after forty years of treating all kinds of patients—from those with autism, to those with severe brain injuries, to those from abusive homes, to those who needed a tumor removed that took with it nearly half of the patient's brain—that anyone can *recover* their brain function and restore their health and well-being.

> Whatever the circumstances of a child's early life, and whatever the history and current state of that child, every human has the built-in power to improve, to change for the better, to significantly restore and often to recover. Tomorrow, that person you see in the mirror can be a stronger, more capable, livelier, more powerfully centered, and still-growing person.[24]

This crystal-clear statement from one of the world's leading experts on neuroplasticity is the exact opposite of what the bullying paradigm wants us to believe. The bullying paradigm has developed a mindset that seeks errors, deficits, differences, struggles, and failures. The bullying paradigm is on high alert to see if it can identify something that allows it to unleash ignorant, at best, and unempathic, at worst, labels. It operates from an indoctrinated, rigid belief system that the brain is fixed and cannot grow. While it has no research to back it up, it passionately adheres to its own principles, in the same way that someone might believe in a cult.

There was extraordinary potential in Angus's brain that even in 2008 the school system failed to understand, let alone assess. Angus was making up answers on the test because he couldn't remember anything from the book he had "read" with his eyes. He had a visual processing deficit, right? Partially right. The test in third grade was measuring our child's ability to recollect information and reproduce it. That's pretty limited when you think about it. The test failed to recognize that our son was making up the answers because he possessed the extraordinary potential required to be a creative writer. He wasn't actually going to read, record, and reproduce other peoples' fictions, which is what the test was designed to measure. Instead, our son would become an adult who makes up his own ideas, stories, and fictions. At eight years old, Angus was showing his teacher, and us, that his passion and talent lay in creative writing, but this extraordinary potential was locked away in his mind. He's now twenty and studies creative writing. He is in the midst of writing multiple intertwined and sophisticated original fictions. Career counselors frequently ask their clients to go back to that eight-year-old time in their lives to remember what it was that they loved to do. "What was it that captured your curiosity, ignited your desire to pour time and energy into it, gave you a sense of purpose and happiness?" Think back to *your* eight-year-old mind and brain because in that breathtaking galaxy lies a constellation of present-day wholeness, alignment, purpose, and joy. Our son was destined to be a writer of culture. How many of our brains have their potential locked away because we are not factoring in neuroscientific insights across the education system and we are not creating tests that are tailored to individual brains? And worst of all, we are generating inaccurate labels and belief systems that trap individuals in the cage of learned helplessness.

# Step 4: Unlearn and Rewire

Cognitive deficits can be rewired in the brain and so can the neurological scars or deficits caused by bullying and abuse. It's empowering to fully realize that with repeated, deliberate practice, your Mind-Brain-Body can be shaped by you and not by past trauma. When you've been bullied or abused, it's vital to separate out the bullying narrative from your authentic self. All too often, the obstacle to repairing the harm of bullying and abuse is that we mix up the story or account that has been projected onto us by bullying or abuse with our own story. We struggle to separate out the two versions, and they become entangled in the brain.[25] As a reader of culture in the bullying paradigm, you might tell yourself a story about how you're weak, lazy, spoiled, frigid, a snowflake, a retard, or a pu**y. You might blame and shame yourself and believe that everything's your own fault. You may isolate yourself from others. Blaming, shaming, and ostracizing are behaviors that serve the outdated, bullying paradigm. They can be replaced in the *neuroparadigm* with: celebrating mistakes because that's how the brain learns, encouraging growth mindset because that's how the brain sails over obstacles, collaborating and connecting because the brain flourishes with social investment. The goal in this action step is to become more deliberate in your rite of passage from being a reader of the bullying paradigm to being a writer of a new paradigm, one that factors in the brain's crucial role rather than ignores it. The goal is to exit the cage of learned helplessness.

Professor of social work and management Brené Brown explains that in navigation, "the term reckoning, as in *dead reckoning*, is the process of calculating where you are." She adds that the original meaning of reckoning was "to narrate or make an account."[26] Because we mix up the abusive story with our authentic story, we cannot seem to get our bearings. We never seem able to figure out where we are, let alone who we are. This action step is designed to help you navigate back to agency, freedom, and wholeness. The ultimate goal is to stop sacrificing Mind-Brain-Body to the bullying paradigm's siren song of ignorance, quick fixes, and destructive self-medication. The reckoning is an important first step because it asks you to be conscious of a story that has been projected onto you by others that you're ready to let go of. This allows you the opportunity to then rewire or rewrite that story into a new one of your own choosing.

Another way to think about this process is you are aiming to take a clash of voices from Mind-Brain-Body and rewire it into a harmonious dialogue. So to put it into simple terms, let's say your neurological scars originally caused by bullying and abuse lead you to believe and act as if "you'll never be an athlete" or "you're lazy" or "you're pathetic" or "you're stupid." After you've found the courage to hear these statements as foreign, not belonging to you or originating within you, it's time to rewire them. How do you do it? Every single time one of these false mantras enters your consciousness, you stop what you're doing and rewire it. "You'll never be an athlete" is replaced by you skipping on the spot for five minutes, then ten, then fifteen, and so on until you're ready to run a half-marathon. "You're lazy" is replaced by you setting realistic goals and accomplishing them. "You're pathetic" becomes a mindfulness session whereby you consciously tap into your resilience; you ignite the neural network for grit; you remember every single time you demonstrated strength of an emotional, spiritual, ethical, or physical kind. "You're stupid" is extinguished with writing down on a piece of paper every single time you had a creative, original, innovative idea; every single time you made a witty joke; every single time you problem-solved your way out of some sort of impasse or crisis; every single time you actually noticed your brain doing its miraculous work of gathering information, assessing it, combining it, questioning it, putting it into new forms never seen before; every single time you acknowledged one of the eighty-six billion stars shining within your skull. The more you fire up all those neural networks, the more you don't leave room in your brain for the outdated and inaccurate statements about you. Remember, the brain has limited cortical space so there isn't enough room for bullying and abuse in your brain when it's full of energized, movement-oriented, active, positive, and empowering neural networks.[27]

John Arden is the director of training in mental health for Kaiser Permanente. Arden does not dwell on the importance of talking through trauma, rehashing the past, or analyzing the source of destructive or harmful thoughts, feelings, and behaviors. Instead, he reiterates the key mantras of neuroplasticity that we have discussed and need to constantly foreground: "Cells that fire together wire together."[28] In other words, once you've got your bearings, once you've done your reckoning with your past, it might be time to stop firing up those neural networks.

If you were bullied or abused, you can talk about it, analyze it, dissect it, and affix all kinds of present-day beliefs and emotions to it, but this may wire it even more into your brain. The beauty of neuroplasticity is that when you make changes to what you remember and how you remember it, what you do and how you do it, your brain overrides the old neural networks with new ones. Once you've done the initial reckoning, hopefully with a mental health practitioner, then you can move on. Why give more fire to those who bullied or abused you? Why wire them more firmly into your brain? Get rid of them. Recognize that they might have ruled the "old you," but you are letting that outdated self pass away and replacing it with a new self. This is what initiation from being a reader to being a writer is all about. While the old you, the reader of culture in a bullying paradigm, might feel stuck in the cage of learned helplessness, the new you, the writer of culture, acknowledges that you are free.

Research is clear (and extensive, replicated, peer-reviewed, and consensus-building) that how your brain operates, what it defaults to, how it interacts with your mind and body is up to *you*. This knowledge should inspire you, but the transformation takes work. It's not a quick fix. You can't just wake up one morning and say, "Hey, you, yeah you, the ones who bullied and abused me, well, I'm sick and tired of the way you're wired into my brain and I'm going to replace you with new, healthier, more supportive, more fulfilling neural networks." And then poof, you're the new you. That approach doesn't work. It's the same with your body. You can't wake up one day and decide to get in shape, and lo and behold you've got a lean, muscular, flexible physique. You *can* have that body, but it takes daily work. It takes commitment. You have to work out, not quit, believe in yourself, maybe get a personal trainer to advise you. Same with your brain. Bottom line is: the work is worth every single moment because it is ultimately up to you how strong, flexible, and resilient you want your neural networks to be.

Arden offers a telling twist on the oft-quoted "neurons that fire together wire together." He adds the equally important brain principle "neurons that fire apart wire apart."[29] In other words, the less you ignite the neural network for "I was bullied," "I was abused," "I was victimized," the more you're going to erode the connections, links, and networks for it in your brain. You need to acknowledge the neurological scars, the damage done to the brain, the harm. Ignoring it blocks you or a mental health professional from diagnosing it and then putting into place a tailored remedy. However, you don't want to identify with these nonauthentic versions of yourself. You are the writer of your life-story, and those who bullied and abused you were merely interruptions. Very destructive interruptions no doubt, but never let them tell the story about you.

If you take a neuroscientific approach to rewiring your brain, then it requires you to stop remembering, retelling, reiterating, and maybe even repeating the past. It demands that you remember something new. Do something new every single time the default or pattern starts up its weary, broken record about what was done to you. It is completely and utterly within the power of your brain to rewrite your story, rewire your neural networks, and replace the bullying and abuse storylines that are outdated with innovative, startling, change-making new narratives.

Put simply it might go something like: "I was bullied and abused by three teachers in high school. They hurt me. They damaged my brain. I am now the way I am because that happened to me." It's up to me to change this old story into a new one like: "I survived mistreatment by abusive adults in high school. They were psychologically messed-up people. Their misconduct impacted my brain when I was young. But they have *no* hold on me now. I've worked hard to rewire my neural networks to fire up on self-empathy and self-compassion." Note, I didn't skip the reckoning. I acknowledged the harm done, but it leaves no legacy or lasting harm unless I let it. The key to your brain is remembering that you have neuroplasticity, which means you have agency. Your choices, decisions, and conduct shape your brain much more powerfully than what was done to you in the past. You have the capacity in your brain to let go of negative neural networks and rewire them into positive ones.

Arden puts this in a straightforward way: "Neuroplasticity illustrates the phase 'Use it or lose it.' When you use the synaptic connections that represent a skill, you strengthen them, and when you let the skill lie dormant, you weaken those connections. It's similar to the way that your muscles will weaken if you stop exercising."[30] What's the best way to transform those who bullied or abused you into weak, flabby muscles? Don't use the neural networks or paths that lead you back to remembering your adversity or suffering. The second your brain wants to march down that path, you need to activate an inner rebellion. Gather your guerrilla forces and launch a coup. This is where your mind is key. It's the leader of the rebellion because, while your brain constantly circles around bullying and abusive past moments, worrying that they might hurt you, because it struggles to tell the difference between past and present, your mind's job is to make clear that that was *yesterday*. It happened a long time ago or an hour ago. Regardless, it's over and your mind is back in the driver's seat. Your mind needs to consciously and kindly (reserving a bit of curiosity in case there's important info it needs from the trauma recollection) shift you down a different path. Your brain may resist because not only do brains worry excessively; they are also not naturally trailblazers. Brains prefer well-trod paths and put up a fuss when asked to pull out the machete and carve a new route. As Rick Hanson has elucidated in his work, the brain is primed by evolution to vividly remember harmful moments because that's how it helps you survive. At a certain point, however, when you have established safety, you need to forget people and events that are dangerous and hurt you and replace them with people and events that are affirming and heal you. He refers to this process as "brain building" and explains that there are five key components to building a stronger brain: duration, intensity, multimodality, novelty, and personal relevance.[31]

- First, you need to put in time and effort. It's not a quick fix. *Duration* is key. Just like when you practice typing or ping-pong or the cello, it takes time, energy, effort, perseverance, and faith that your hard work will pay off.
- Second, you need to go at the practicing with *intensity*. Take all the passion you have and channel it into rewiring your feelings of helplessness, depression, and futility into feelings of independence, joy, and hope.
- Third, bring a mindful mindset to your practice. When you exercise your power, your lack of helplessness, then become aware of all the ways that you can experience it. Apply *multimodality* to your practice: feel power in your

body, take on power poses, say it out loud, sing it, write it, draw it, share it with a friend or your tribe, visualize it. The more modalities or routes that you take to walk out of the cage of learned helplessness, the better.

- Fourth, surprise your brain. Tap into *novelty*. Don't make your experience of independence, joy, and power be routine and boring. The brain is far more likely to wire something in that is fresh and new.
- Fifth and finally, commit to deliberate practice in a way that speaks to your unique brain. Don't try to do exercises in gaining freedom, confidence, and bliss according to someone else's ideas about what to do and how to do it. Instead, make exiting the cage of learned helplessness something that has *personal relevance* to you.

It might feel challenging. You might feel overwhelmed. But it's actually possible to rewire your neural networks by practicing. The key is you need to do it frequently and consistently. As Arden puts it: "Repetition rewires the brain and breeds habits."[32] Bad habits become more entrenched every time you repeat them. Every time you have a thought or do a behavior that you are tired of or that harms you, stop and make yourself conscious of the fact that you are once again rewiring your brain in that negative way. As Doidge explains:

> Sometimes one can manipulate the use-it-or-lose-it principle to undo brain connections that are not helpful, because neurons that fire apart wire apart. Suppose a person has formed a bad habit of eating whenever he is emotionally upset, associating the pleasure of eating food with the dulling of emotional pain; breaking the habit will require learning to dissociate the two.[33]

In short, you've got the power to rewire your brain in a different way. You can daily harness the power of your brain to *not* use the limiting belief or unhealthy habit and thereby lose it over time. By separating or distancing positives (eating) with negatives (trying to dull sadness, fear, anxiety, whatever), you dissociate them (unwire them) in your brain. Remember, it takes time, effort, and energy. So if it doesn't happen right away, that's to be expected. While it takes a lot of time to develop a destructive habit, it also takes a lot of time to replace it with a constructive one. In fact, it's twice as hard because learning is much easier than unlearning.

While there isn't a quick fix, there is a magic bullet to this whole rewire-your-brain business. Unfortunately, the magic bullet has one of those long-winded neuroscientific names: brain-derived neurotrophic factor. Fortunately, scientists who talk casually using these kinds of terms have come up with an acronym: BDNF. The first time I read about BDNF was in John Ratey's book *Spark*, where he shares that BDNF is like a miraculous fertilizer for neuroplasticity, as well as the birth and growth of new brain cells.[34] Guess what bullying and abuse do to BDNF in your brain? You guessed it: the chronic stress that all too often accompanies bullying and abuse causes your brain to produce less of this crucial protein for neuroplasticity and neurogenesis. Research shows that victims' brains produce less BDNF and the BDNF deficiency reduces the brain's ability to produce enough dopamine, serotonin, and other key chemicals that help you to think clearly and enjoy emotional stability.[35] This is exactly why you have got to shut down the bullying and abuse networks and rewire them with empathy and compassion networks.

Merzenich informs me that "BDNF was measured in several brain training studies using BrainHQ," which is a site not only for research but for all individuals to do weekly brain "workouts" that strengthen and fuel the brain, maintaining its overall health and wellness. What did the studies show about BDNF? "It was very substantially elevated by training." If a brain had been traumatized, so that the BDNF was "pathologically low," brain training would return it to normalized levels. Merzenich explains that the key is to "make the most out of effort spent: training your brain in ways that advance the learning-control and learning-enabling machinery. Much of our research over a period of about a decade targeted these issues."[36] The research is clear. All we need to do now is start training our brains according to the neuroscientists' knowledge of how to heal and optimize our brains' performance. With your brain veritably defined by its dynamic neuroplasticity, the cage that holds you back is an illusion. Your brain potential is unlimited.

# CHAPTER 5

# Prevent the Cancerous Confusion of Bullycide

## *Step 5: Grieve*

In 2015, in my second year at a new school, I taught a brilliant girl in an International Baccalaureate English Literature course. To protect her privacy, I will call her "Ellen." She was an international student from China who boarded in the "family stay" program at the private school where I was teaching. English was Ellen's second language; nonetheless, she produced insightful and articulate work. A slender, willowy girl, she had shoulder-length black hair, expressive eyes, and a wry smile. In March 2015, when the story about the abuse at my previous school broke in the *Toronto Star* and on CTV's *W5* program, she said she was proud of me for speaking up on behalf of students' rights to attend school safe from abuse. As the year wound to its conclusion, the final assignment was for students to write a personal essay in preparation for a provincial exam, and Ellen wrote about suicide. It caught me off guard because she was so smart and funny and had wonderful friends. I took her essay to one of the school's counselors because it worried me. Ellen would leave class sometimes, apparently so anxious she couldn't bear to stay, and she would go to the counselor's office as a safe place to calm down. She said it was because she couldn't stand English, which was meant to be funny, but it wasn't anymore. I left the essay with the counselor as she knew far more about Ellen's mental health than I did.

In a final meeting before summer, faculty were informed that Ellen would be required to live with one of her parents if she were to remain at the school for her final year. She had been living with the school principal and his wife as "family stay" parents while her parents were in Macau, but due to her compromised mental health, it was deemed necessary for her to live with family. We learned that her father would be coming to stay with her for twelfth grade. Perhaps this would provide the stability she needed. From what I had been told, Ellen had been in boarding schools since she was nine, which struck me as awful, but that was perhaps my cultural bias. Alex Renton, who was put into boarding school at the age of eight, condemns the British elite for sending children off to boarding school especially when they are so young. He suffered this "attachment fragmentation" with his parents, as do many others.[1]

I didn't teach Ellen the following year, but I would see her from time to time. She asked me to go out for tea at the start of the year with another of her friends and regaled us with hilarious stories of her escapades. I didn't forget about her essay, and there was something concerning about her stories. Possessed of a clever wit, constructed on a deep intelligence, she always had something funny to say, but through the irony or sarcasm, she seemed anxious. In October 2015, she came up to me in the library and asked if we could talk. She informed me she was being sexually harassed by a teacher. And then she told me who it was. It wasn't just a teacher; it was the school's principal. The principal who hired me knowing I was a whistleblower on child abuse at the other school. The principal who was so kind to my son. The principal who had been Ellen's family-stay "father."

I didn't know then what I know now. Those who abuse are adept at presenting as if they would be the last person on the planet to harm anyone, let alone a vulnerable child. Harvey Weinstein was a pioneering supporter of women directors and women's issues, when he wasn't sexually abusing them. Jerry Sandusky ran the award-winning "Second Mile" charity for orphans, when he wasn't raping boys in the shower at Penn State University. Larry Nassar was known as a doctor who would do *anything* for his patients, when he wasn't abusing hundreds of them. One of the teachers who abused Montgomery very *publicly* supported a disabled former student, while *privately* he slammed the boys with homophobic slurs and watched the other teacher call them "f***ing retards." Here we are again face-to-face with Dr. Jekyll and Mr. Hyde, and yet we seem to fall for it every time. When Ellen told me that the teacher who was sexually harassing her was the principal, I should not have been surprised.

Regardless of my shock, I had a legal duty to report, just as I did at the other school. The problem was I no longer trusted the Teacher Regulation Branch and the Commissioner, but since this was a report of sexual abuse, I hoped the police could take charge. Although the headmaster at the other school reported the multiple allegations of emotional and physical abuse to the police, they couldn't do anything because they said the abuse was not criminal in nature. When we asked about the hundreds of assaults done to Montgomery where he was yelled at in the face in threatening scenes of public humiliation and then grabbed and detained for more when he tried to get away, the police officer said that a judge would say the teacher was simply "motivating" him. A boss or manager would be charged with assault if they behaved this way to an employee on a single occasion, but the rules change when the power imbalance is between an adult and child. Assault morphs into "motivation." Still, the police can usually intervene when the abuse is sexual in nature, so I held some small hope Ellen would be protected.

I informed a school counselor what Ellen had told me. Together we reported to the administration. The principal was suspended. The administrator asked me to take Ellen's testimony, which I did, just like I did with eight students at the other school. The police were contacted, and they began an investigation. I would have liked to wash my hands of the whole sordid repetition of my traumatizing experience at the other school, but Ellen was suffering terribly and sending me e-mails about her fear, her relief in reporting, her guilt about how the principal would suffer, and her self-loathing, which I struggled to understand. While she was clear about the sexual harassment in October, as the year unfolded, she seemed to become more and more confused. I had already been through the inner workings of the bullying paradigm, and so I was even more frustrated at how much she seemed to fault herself for what happened. It was textbook victim blaming, but she was blaming herself. She worried that maybe she had just been "too sensitive."

While the principal never returned to the school, everyone was told a lie about his whereabouts. Of course this would fuel Ellen's confusion. She was told that she could participate in restorative justice, and if so, then the principal would not be charged. Why would she even be asked to protect him? Support him and ensure he was not held publicly accountable? She was a teenager, and he was a school leader in his sixties. Surely, his victim did *not* need to be included in how the law dealt with his abuse. None of it made sense to me, and I simply answered Ellen's e-mails to the best of my ability. I was told she had a lot of excellent mental health support, and not being an expert myself, I tried to just keep her clear that she had done the right thing. At one point, with all the encouragement to do restorative justice, I asked her if she had a lawyer. She said no. It seemed like a pretty one-sided battle to me. Children lack the knowledge and experience to keep themselves safe, but when the system is designed to protect the adult perpetrator, kids don't have a chance. An international boarding student—whose family speaks a different language, belongs to a different culture, and follows a different set of laws—is at particular risk.

In the spring, the police dropped charges against the principal. Ellen overdosed and ended up in the hospital. Her e-mails were making me frantic. I felt completely powerless, and all I could do was write the Ombudsman's Office and tell them what was happening and how worrisome it was. The school community and faculty had been told that the principal was on "sick leave"; then months later the administrators changed the lie to "stress leave." Why cover up what he had done? If he was innocent, he'd be exonerated and back at school.

No wonder Ellen was so confused and felt guilty. The oblique messaging from the bullying paradigm to victims is that it is their fault. They usually have a loyalty-bind with their abuser and blame themselves for the adult's suffering. When their world reflects to them a need to cover up what the adult has done, it sends distorting and harmful signals to the brain.

Ellen wrote her International Baccalaureate exams in May. These very challenging exams give students credit for the first year of university. Ellen scored at the top of the whole class. Directly after, she attempted suicide again and was locked up in the psych ward at the hospital for a couple of weeks to keep her safe and get her back to health. She asked for me to visit, and I saw her there. I was afraid and distraught at her desperation. My legs shook uncontrollably as I sat at her bedside. I tried to get her to understand that if she hurt herself, she would be hurting so many of us that cared for her. By this point, I was struggling. I found myself starting to shake whenever I went to the school, with tears welling up in my eyes at hypocritical meeting after meeting, as administrators and faculty geared up for the end of the year and graduation. I felt like I was as confused as Ellen and starting to fall apart. I requested stress leave so that I would not have to go to graduation and pretend that everything was normal.

After the graduation ceremony, I heard from Ellen, who was in a fury. Right in front of her, without warning her, at *her* graduation, the school administrators put on a celebration for the principal. No one seemed to question why he wasn't in attendance at the ceremony. The administrators talked about his eighteen years of service and what a wonderful contribution he had made to the school. Ellen said people were talking about how he dedicated so much to the students that he had a kind of breakdown and had to go on stress leave. All of the inertia on the part of the Ombudsman's Office, the last resort for child protection at this stage, and the hypocrisy on the part of the school administrators was too much for me. I had to realize that the bullying paradigm did not only rule at my previous school; it was also in other schools. It was a prevailing, entrenched belief system that could not be toppled with distressed pleas to protect kids.

Like Montgomery, Ellen went off to university much to my relief. I felt that if they could just put the abuse into "the rear-view mirror" as my son would say, then they could heal. They could forget and move on with their lives. Ellen was a gifted science student and wanted to become a doctor. She made it onto the varsity synchronized swim team. She seemed so much better despite the occasional e-mail that give me an electrifying jolt of fear. I always wrote back to remind her she had done the right thing. *She* was the victim. It was *not* her responsibility to protect the abusive principal. I don't know if she listened to me or could even hear my words. She wrote in late summer, before her second year at university, to see if I could have tea with her, but I was out of town. I did not know anything about what she posted a month earlier on Facebook. It was dated July 5, 2017, and the title was "Started New Job at DF medical beauty centre."

Another student showed it to me in December, five months after Ellen posted it. I read:

> I have been thinking about posting for a while. It is definitely a tough decision to make. Even if only you are a random friend I made in some random lecture, I still invite you to read the following. Of course it's going to be a long post. For those who don't care, please scroll down and ignore this.
>
> For years I have been struggling with mental health problems. I have lots of diagnoses; borderline personality disorder, depression, anxiety, primary obsessional OCD, dissociation disorders. In high school I even struggled with eating disorders due to anxiety. I have a history of sexual abuses and some family problems. My depression problem started since I was in elementary school. I didn't know it until grade 11. Not long after I started receiving treatment, one teacher from my high school sexually harassed me. He betrayed my trust. Since then everything went out of control. I made several attempts to try to end my life. For a while I was hospitalized on a monthly basis.
>
> The point of telling you this is not to try to get empathy from you. When I am out with people, I always laugh, smile, joke around and make everyone laugh, but deep inside, I can barely feel that emotion. Not that I'm pretending to be happy. I do try my best, but due to the traumatic events, my body shuts down my emotions as self-defence. The only emotions I can feel are deep depression and anxiety. I always feel so insecure. My relationship with people can be very short and intense. Sometimes I might say things that sound attention seeking and manipulative. I do self-harming. For most of the time I feel like there is a colourless wall that separates my physical body and my mentality. Intense but short episodes of anxiety and depression. These are all symptoms of borderline. There are lots of stigma about borderline personality disorder. All I can say is that as someone who suffers from it, it's pain that can't be translated into words.
>
> Sometimes when I fall into the darkness of depression, I look back and wonder what would it be like if I wasn't discovered after I made those attempts. If I had succeeded, most of the people around me probably wouldn't even notice that I had been struggling. If I was one of those people I would definitely feel some guilt in me. I would definitely think back and wonder what could I have done differently to try to save a life. In addition, some people who actually try to help can't help because they don't understand what it feels like to be depressed and to feel like there is no way out. So I wonder maybe sharing my story would at least give people a chance if they want to help.

I am still struggling. I am still at risk of taking my life. I can't help it. I have been trying so hard and things haven't changed for years. I understand that sharing this means a lot. I worry about being stigmatized, but at the same time if this is a disease, why can't people treat it as a disease?

Just like people with cancer, we who suffer from severe depression are also at risk of dying. Just like cancer, we can't control how this disease goes. Just like cancer, we can't just wave a magic wand, smile, and the disease would disappear. Just like cancer, deep inside we want to live, but we can't control the disease.

If you read till the end, I invite you to take some time to learn about mental illness. There are many people out there who are suffering, and maybe just asking them how they are doing can help a lot. I was once saved by someone who was willing to listen to me and that I told her my plan.

Thank you for taking your time to read my verbal diarrhoea.[2]

As I sat in a café with my former student, her phone extended so I could read what Ellen wrote, tears fell onto the table. Still only a teenager, Ellen was yet able to clearly articulate what is dangerously wrong in a world that operates according to a bullying paradigm. Her clarity and expression of suffering were powerful, and yet the world doesn't seem able to hear it. Ellen's posting on Facebook exposes the anomalies, the fundamental flaws in the outdated paradigm that behaves as if the illness in the brain is not relevant and does not need to be healed.

At the outset, Ellen discloses that sharing her mental health saga was a "tough decision" to make. She lists the "diagnoses" that one would assume led to a series of remedies, but the list is long and there's no moment in the posting to suggest she ever received any effective treatment. This is not a criticism of mental health professionals, just like questioning the approach of educators or parents throughout this book is not to blame them. The overarching goal is to ask some tough questions around why we work around the clock, with a great deal of attention and visibility, on a cure for cancer, but we do not seem as invested or public when it comes to disorders or illnesses in the brain. In fact, the moment anyone mentions mental health, we instantly have to talk about the "stigma." Why? Why is it acceptable to have something wrong or injured in the body but not in the brain? And to even suggest the brain and body are distinct, or to privilege one over the other in terms of care or intervention, does not make any sense. Neuroscientists ask questions like "How does being victimized lead to emotional disorders and suicidality?" Their response is that bullying has the power to alter "the physiological response to stress" and, most importantly for Ellen, "affect the telomere length or the epigenome, by interacting with a genetic vulnerability to emotional disorders, or by changing cognitive responses to threatening situations."[3] Ellen's vulnerability due to past abuse appeared to inform her clearly disordered cognitive responses to the threat of sexual harassment by a powerful male figure. Even more concerning, telomeres—a compound structure at the end of a chromosome—when lengthened are directly correlated with cancer.[4] Her metaphor that what she is feeling compares to cancer is *not* a metaphor. It's closer to a medical diagnosis.

Although Ellen does not refer to research, she does draw a connection between sexual abuse in childhood, family problems, and the long list of labels that have been attached to her. Then she issues a very clear-minded statement: "One teacher from my high school sexually harassed me. He betrayed my trust. Since then everything went out of control. I made several attempts to try to end my life. For a while I was hospitalized on a monthly basis." Just as she was getting some kind of support and healing for "depression," once again she was pulled into a sexual abuse scenario. It seems like the final straw, as now "everything went out of control." Note that the harm does not occur to her body. It is a betrayal that scrambles her brain. It confuses her understanding about who to trust and who to fear, who is supportive and who is dangerous. Her faith in empowered adults gets shattered: "He betrayed my trust." She correlates this blow to her brain—which is not a metaphorical injury as documented in extensive research—with suicidal ideation: "I made several attempts to end my life." While she's frequently hospitalized, it appears the medical experts cannot heal her sense of betrayal, her profound confusion, or what it has done to her brain. Month after month, they struggle to stop her impulse to destroy herself. Her brain does not appear to be mentioned, let alone studied. Shouldn't the experts be wondering what is happening in her brain that has thrown some kind of switch, so that instead of fulfilling its deepest impulse to survive, it is now trying to enact its own death? Merzenich informs me that Ellen appears to be suffering from "the dysregulation of serotonin and the change in the balance of emotional valence." Few of us are taught to factor in the way in which our emotions are connected to and influenced by valence or chemical balance in the brain. Serotonin is a neurotransmitter that has many multifaceted functions in the brain including the regulation of mood. Ellen's brain suffers from a mood imbalance. Her brain can no longer balance happiness and sadness. In fact, as Merzenich notes, her whole emotional self is "certainly tilted toward the dark side, toward sadness for Ellen."

In response to my frustration that Ellen does not get the kind of diagnoses and intervention she obviously needs for her brain, Merzenich concurs that she apparently has a "neurological distortion," which he says is "easily detectible and easy to impact with brain training." If there is any part of this book that is crucial, it is that key statement. For yourself or anyone you know who is suffering from profound "sadness" or depression, namely serotonin dysregulation, you and they need to know that Merzenich and his team's research has shown in animal models that "serotonin production and release can be dramatically impacted by intensive brain training."[5] In chapter 6, we will discuss this in more detail.

Ellen continues to be very clear that she has been harmed. She articulates the way in which trauma upon trauma has divided her. The public self smiles and laughs while the private self feels depression, anxiety, and insecurity. Note again her brain's confusion about how to judge emotional valence. She explains: "I do try my best, but due to the traumatic events, my body shuts down my emotions as self-defence." In psychiatry, what she's describing is alexithymia, which is the Greek term to describe "not having words for feelings."[6] Ellen makes a mistake here in believing that it's her "body" that shuts down her feelings. It's not; it's her brain.[7] It's a neurological distortion. Countless neuroscientific studies show that when an individual has suffered trauma, especially from abuse, the brain remains on high alert, waiting for the next blow. The parts of the brain engaged in survival and safety take up most of the brain's energy and focus, overriding other brain functions. The brain develops hypervigilance to survive.

Ellen describes an invisible, "colourless wall" that divides her physicality and her mentality. Ellen feels that she is becoming two different selves: one self is the body and the other self is the brain. Expert labels strive to capture this phenomenon: borderline personality disorder or dissociation disorder. It is becoming a battle within her as she develops a Mind-Bully and attacks her own self as if this part of her is someone else, another "personality." How else to make sense of eating disorders and self-harm whereby an individual becomes the aggressor who attacks and harms their own self? So many of these behaviors, of course, are correlated with abuse. One of the ways children survive the abuse being done to them by adults in positions of trust and power is to dissociate. Their brain cannot cope with the cognitive dissonance. The harm is less traumatizing if it's happening to "someone else." That makes it easier on the victim. As trauma expert Bessel van der Kolk explains,

> Traumatized people chronically feel unsafe in their bodies. The past is alive in the form of gnawing interior discomfort. Their bodies are constantly bombarded by visceral warning signs, and, in an attempt to control these processes, they often become expert at ignoring their gut-feelings and numbing awareness of what is played out inside. They learn to hide from themselves.[8]

Ellen tells her reader that dividing herself, being both internalized abuser (or Mind-Bully) and suffering victim, is inexpressibly painful. There are no words for it, and yet she vigilantly tries to explain. She tries to push through the stigma and shame.

Frequently, children do not get the protection they need to prevent abuse, nor do they get protection from abuse, even when they report. Frequently, they are told they misunderstood or it couldn't be true or that the adult's reputation may get harmed or that they caused the abuse. Children cannot protect themselves; the adults in their world fail to protect them; they internalize these mind-bending flaws at the heart of the bullying paradigm and find themselves full of shame and self-loathing. Instead of clearly seeing the abusive individual as the one doing the harm, they believe they are the "monstrous" one. This isn't new research. In the 1980s, psychiatrist Roland Summit and psychiatrist Alice Miller articulated it in detail in their work.[9] Research by neuroscientists confirms psychiatric insights from forty years ago about what abuse, and the failure to stop it, does to the brains of victims. However, not even decades of neuroscientific research and the glaring evidence on brain scans has had the power to topple the bullying paradigm's construct. What will it take for society to exit this destructive, outdated paradigm and enter into a new way of being informed by research rather than entrenched beliefs?

Ellen says her goal is not to elicit empathy, and in fact, she makes clear in this section that her goal is to shine a spotlight on the failed approach of the bullying paradigm:

> Sometimes when I fall into the darkness of depression, I look back and wonder what would it be like if I wasn't discovered after I made those attempts. If I had succeeded, most of the people around me probably wouldn't even notice that I had been struggling. If I was one of those people, I would definitely feel some guilt in me. I would definitely think back and wonder what could I have done differently to try to save a life.

At her lowest, most depressed state, she imagines her own successful suicide and realizes that "most of the people around me" would fail to see her suicide as a bullycide; namely, a murderous act against her own self as a result of her struggle with abuse. The word "bullycide" captures in an intense single concept what Ellen exposes. When you kill a man, it's "homicide." When you kill a mother, it's "matricide," and when you kill yourself, it's "suicide." When you commit "bullycide," you are killing the "bully" who has abused, harassed, and made you so miserable you want to die. However, your brain has so effectively identified with the aggressor, your brain has so profoundly internalized this Mind-Bully, that you end up killing yourself. Isn't this the natural conclusion of victim blaming? Internalizing the harm done begins to unravel their brains. They start to believe that they deserve the abuse. They deserve to die. It is a neurological distortion, but no one is looking at their brains.

Ellen's body looked fine. She was smiling, and so those around her thought she was safe. The teacher who was sexually harassing her was removed and barred from returning to the school. Therefore, Ellen must be safe. What is being missed is that where he harmed her was not her body. It was in her brain, and that injury is not being noticed, let alone healed. Ellen doesn't want empathy for herself, but in the following statement, she again is clear about why she makes the tough decision to post publicly about her mental health struggle. She says that if *she* was one of the people around her, who failed to notice she was fighting for her life, *she* would feel guilty. She writes:

> I would definitely think back and wonder what could I have done differently to try to save a life. In addition, some people who actually try to help can't help because they don't understand what it feels like to be depressed and to feel like there is no way out. So I wonder maybe sharing my story would at least give people a chance if they want to help.

What can we do differently, so that a victim like Ellen is not struggling alone and is not being labeled with all kinds of "diagnoses" that somehow carry a stigma or mark of disgrace? She says we need to "think back" and "wonder." The failure in all those who were around her, myself included, is that we didn't understand her depression and we didn't know that she was trapped in it. And here's the saddest part. Ellen shares this message publicly because she wants to "at least give people a chance if they want to help." She's giving her readers a chance. It's not too late to help her. It's as if she knows there's a way to heal and restore her health and happiness if only we could understand "what it feels like to be depressed." If only she had been aware of Merzenich's research and the brain training that heals neurological distortions.

In the following section, she describes depression in elucidating terms so that we just might understand and maybe even want to help. Her anguish is palpable. Again, she expresses the fear of being shamed or stigmatized for having suicidal ideation. She recognizes that what she suffers from is as much a disease as cancer is. What she can't understand is why the illness in her brain is somehow seen as a moral flaw and differs from an illness in her body. Why is "cancer" of the brain the patient's fault and cancer in the body is not? Why is one addressed morally while the other gets medical intervention? It begs the question: if we treated Ellen's suicidal ideation as a medical crisis within her brain, wouldn't we be more successful with treatment? She issues a cry for understanding and for help: "Just like cancer, deep inside we want to live, but we can't control the disease." This is a frightening statement because her brain has apparently lost control.

Bessel van der Kolk says this is a natural reaction to abuse: "Trauma robs you of the feeling that you are in charge of yourself, of what I will call self-leadership."[10] The leader within Ellen has been usurped by a life-threatening abuser, a Mind-Bully who pushes her to self-harm, suffer disordered eating, hide her true feelings, try to please others while her own world is collapsing, and, most of all, keep secret "the ways [she] has managed to survive" as the stigma is so unbearable.[11] Her brain is aligning or identifying so intensely with this usurper, this aggressor, this Mind-Bully that not only her self-leadership is at risk—her body might be harmed.

Van der Kolk asks his colleagues: "How do we treat people who are coping with the fall-out of abuse, betrayal and abandonment when we are forced to diagnose them with depression, panic disorder, bipolar illness, or borderline personality, which do not really address what they are coping with?"[12] Ellen is clear that her brain goes "out of control" when she realizes that she has been betrayed by a trusted adult. She is told by the medical community that she has depression and suffers from a borderline personality. She does not connect the abuse done to her with the scars on her brain. She becomes consumed with misplaced guilt and self-hate. The school community and educational authorities fail to make clear the harm done by the principal and fail to hold him publicly accountable. Instead, they cover up what he has done in what appears to be a misplaced impulse to protect his reputation and the school's reputation, rather than protect the child he victimized.

Van der Kolk documents in his book on trauma, *The Body Keeps the Score*, the many instances whereby patients turn against themselves, as a way to cope with childhood adversity. He explains: "Rage that has nowhere to go is redirected against the self, in the form of depression, self-hatred, and self-destructive actions."[13] These three descriptions apply to Ellen: depression, self-hatred, and suicide or self-destructive actions. Victims usually have a loyalty-bind to the one who abuses them, especially if it's a family member or other adult in a position of trust and authority. Children are "programmed to be fundamentally loyal to their caretakers, even if they are abused by them."[14] This is true with family members, doctors, teachers, and coaches, whom children are taught to obey, respect, and trust. Children are ill-prepared by society to question, let alone report, on their caretakers.

This loyalty-bind or impulse to protect the abuser is not exclusive to children: "Hostages have put up bail for their captors, expressed a wish to marry them, or had sexual relations with them; victims of domestic violence often cover up for their abusers."[15] Many individuals, reading abuse reports in the media, do not realize that when individuals defend the abusive individual, reported on by others, it may well be that they have a loyalty-bind and have normalized or edited out the harmful conduct. It is extensively documented in research that our brains do not see what's right in front of them, referred to as "inattentional blindness." Moreover, our brain "filters out information that it does not deem pertinent."[16] We have "selective perception," and years of positive encounters with an individual, reported on as abusive, overwhelm the brain, which is being told to see the individual in a completely different way. As Shawn Achor puts it: "We see what we look for, and we miss the rest."[17]

As Ellen discovers, she's expected to support the principal as he is processed through the justice system. The school community is encouraged through falsehoods to feel sorry for the hardworking, long-serving, and dedicated principal who is sick and stressed and must retire early. If he's not to blame, if he's not held accountable, who is? Ellen turns the guilt and hate onto herself. She blames herself. It's a particularly perverse form of victim blaming, but it is common. Speaking up is extremely difficult to do. Trauma paralyzes the victim's memory and ability to coherently articulate the abuse. When your brain is subjected to a toxic or traumatizing environment ruled by betrayal, danger, threat, and fear, memories become disconnected; neuronal networks do not form in the same ways that they do in healthy, supportive environments. Ellen is not publicly acknowledged for her strength, courage, and integrity for speaking up. She is not celebrated at graduation. The principal is.

Ellen is a patient with a life-threatening condition like cancer. Throughout her post, she makes it clear that the cancer developed within her not from smoking or other at-risk behaviors; the cancer emerged because of the abuse she suffered repeatedly as a child. The correlation between abuse in childhood and midlife chronic disease is in fact documented in extensive research, as we will examine in the next chapter. It's hard to know if Ellen knew the research, but there is no doubt that she is aware on a cellular, if not a scientific, level that there is an indisputable correlation between childhood abuse and disease in midlife such as cancer.

Ellen's final words are the most disturbing and concerning because they show she has fallen into the grip of the neurological distortion. One minute, she is in control and again reminds her readers that they can help: "I was once saved by someone who was willing to listen to me and that I told her my plan." If we are willing to listen to her, she may well share her plans to commit "bullycide" and oust the abuser, the harasser, the Mind-Bully who has taken over her brain. But then her tone switches to a self-demeaning one, whereby she discounts her remarkable post and calls it "my verbal diarrhoea." Her articulation of suffering that causes her to divide; the lack of understanding, even of those close to her and medical professionals; the anomalies in the bullying paradigm that fail to take account of the wounds to her brain; her impassioned description of depression, coupled with a cry for help, all collapse into the ultimate insult: "verbal diarrhoea."

Her courageous, articulate, heartbreaking post is abruptly reduced to an ugly, degrading image. Her facility with language suddenly becomes a stereotypical put-down, an insult taken from the bullying playbook. It's the words that a Mind-Bully might use to demean the authentic self. It sounds like the internalized aggressor who makes the victim feel so full of self-loathing that she cannot bear to be another minute on the planet. Ellen anticipates and comes out ahead of the attacks she may well have to endure for speaking up about abuse and daring to link it to mental illness.

Remember, your brain is unique. There is no simple cause and effect between bullying, harassing, abusing, and damage to the brain. Martin Teicher's research findings show that as a child grows and develops, "different brain regions are affected at different times." In other words, abuse and maltreatment impact individual brains in distinct ways depending on the time frame. The effects of "exposure to abuse and neglect are complex," explains Teicher.[18] Those who want to dismiss and deny the damaging force of abuse and neglect will be quick to say that one victim was fine, so if the other one wasn't, it's because he or she was "too sensitive." If you've believed that it's your weakness or your excessive sensitivity that made you have a negative reaction, perhaps long-term reaction, to bullying or abuse, it's not. It just happens to be your unique brain and its signature style of reacting and responding to abuse. It may have depended on *when* the harassing, bullying, or abusing occurred and what kind of maltreatment you'd suffered in the past that it built on or triggered. Extensive, peer-reviewed, replicated research demonstrates that the harm done by abuse and neglect "can vary markedly from individual to individual."[19]

To give you a broad overview and demonstrate the complexity, research shows that when parents verbally abuse their children, it can cause "alterations in gray matter volume in the auditory cortex." Research shows that girls who were sexually abused are left with brain regions associated "with thinning of the somatosensory cortex representing clitoris and surrounding genital area." Child maltreatment scrambles the brain's "sensory processing systems." Child abuse in all forms structurally and functionally alters key components in the brain's "threat detection and response circuit." Child abuse and neglect impact "brain morphology, function and network architecture."[20] Ten years ago, Emily Anthes highlighted Teicher's research in the *Boston Globe*. The research I have just cited from him and his team at Harvard is from 2017. When will this research reach us in the general public, and not only protect us from normalized abuse, but also help us learn how to heal our brains if it's already been done? When will this research save children like Ellen?

Psychological, physical, and sexual trauma in childhood may result in psychiatric disorders that show up in childhood, like they did for Ellen, or later in life, like they do for many others. The damage done may translate into anger, shame, and despair that the victim often turns on him- or herself as Ellen described. As noted, research shows that some victims like Ellen turn the abuse *inward* so that it manifests as "depression, anxiety, suicidal ideation, and post-traumatic stress." Other victims turn their damage *outward* so that it manifests as "aggression, impulsiveness, delinquency, hyperactivity and substance abuse."[21] Too often, these disruptive behaviors, when they occur in childhood, result in further discipline and punishment in homes, in the school system, in the criminal justice system, or in the workplace, which only serves to reinforce an individual's trauma.[22] In adulthood, perpetrators can be far more sophisticated in keeping these destructive behaviors hidden behind closed doors. We must understand that bullying and abuse are a self-perpetuating cycle.

When I speak of accountability, I do not mean the perpetrator deserves blaming, shaming, and ostracizing. These are the tenets of the bullying paradigm that are outdated and broken. What matters is that we harness neuroscientific findings to understand that abuse is a cycle: an individual who abuses others was, in most cases, once a victim. Do we heal them or bully them? Do we use a scientific approach to neurological distortion, like we do to cancer, to rehabilitate their hurt brain like their body? If not, why not? Neuroplasticity means we can put every effort into saving someone who has turned abuse on themselves and is suicidal. Neuroplasticity and targeted brain training mean we can put every effort into holding an abusive individual to account, while we can also put every effort into repairing their neurological distortions. Some individuals cannot be rehabilitated, but that doesn't mean we should give up on all of them.

We discussed how the abusive teacher Terence Fletcher in *Whiplash* suffers from "borderline personality disorder." We noted how Martin Teicher uses technology to examine the brain of an abusive teacher like Terence Fletcher only to discover that what makes someone become a split-personality, or a "Jekyll and Hyde," is childhood abuse. To be more precise, we then noted how the medical term that describes a split-personality, or a Dr. Jekyll and Mr. Hyde, is "associative identity disorder." It is when two "seemingly separate people occupy the same body at different times, each with no knowledge of the other." It is a more severe form of "borderline personality disorder."[23] "Borderline personality disorder" is a more accurate way to understand an abuse victim like Ellen who turns against her own self to the point of suicide.

After Ellen gave her testimony to me and the school administrator, I wrote her an e-mail to see how she was feeling. She replied:

> I still feel very guilty for what I have done, even though I know it was the right thing to do and to some extent, I was forced to speak out. He is so nice to me and he cares so much for me. I seriously don't want anything to happen to him.

She felt "very guilty" and wanted to protect him, make sure nothing happened to him? I told her by speaking up, she had likely saved other victims. I tried to help Ellen see that her feelings of care were understandable considering how abuse works, but that they were also misplaced. I explained to her over and over again that it was the *principal* who should have feelings of guilt. I wanted her to see that during the grooming phase of child abuse, the perpetrator uses "being nice" to manipulate and seduce the victim. It was incredibly frustrating and sad to see this child taking on such responsibility in the face of adult betrayal. No matter what I said, I could not seem to break through her belief system. She explained her sense of abandonment as she had pushed through the loyalty-bind and dared to speak up: "I am being left alone to figure out how to cope with all the guilt and hatred I have toward myself, just like an adult." The child who speaks up and does not publicly receive confirmation that she has done the right thing therefore identifies with the aggressor and believes she has committed a wrong. Her response is *not* relief at now being safe. Her response is *not* feeling proud that she found the courage to report the abuse. No, she is rendered "guilty" in the broken system. The failure of adults to tell the truth about what has occurred, adults empowered and responsible for child safety, makes her feel "hatred toward" herself. Again, the Mind-Bully has control of her brain, and she turns all the loathing she should feel toward the principal who betrayed her against herself.

In this e-mail, she's angry about being "left alone" and about being expected to cope "like an adult," which is particularly cruel because the whole crisis has occurred due to her position as dependent child, a child who trusted the principal, leader of her school, credentialed with being a teacher and guide to children, put in the trusted position of being her home-stay "father." Yet, in the broken system, the academic leaders are rushing around trying to protect the principal's and school's reputations, even if it requires covering up abuse. In the bullying paradigm, the faulty belief is that once the body is protected, the child will be fine. Once the pedophile cannot come on campus and access the bodies of children, then safety reigns again. This is completely backward because abuse infects and distorts the brain. Children who are targeted, who have loyalty-binds to the powerful adults in their world, who are wired with empathy to survive, struggle to see clearly that they have been victimized. They fail to report for many reasons, but one of them is that they don't know they've been abused. Some even come to believe that the abusive adult "loves" them.

Ellen almost sees the reversal that was slowly unfolding whereby she would be expected to take on an adult role while the actual adult, a principal in his sixties, would be protected, coddled, and fretted about like he was a hurt child. It was confusing to me and mentally devastating to her. Many in our school community were so worried about him while on "sick leave" and then "stress leave." Many believed he'd worked so hard in his service to students and the school that he'd needed to "retire early." Ellen graduated and went to university. Although her body was safe from the principal, her brain continued to suffer. She could not stop how noisy and hazy her brain had become despite how brilliant it was. She didn't know there were evidence-based ways to heal and repair her neurological distortion. On July 5, 2017, Ellen posted on Facebook asking to be understood, issuing a cry for help.

Betrayed, confused, full of cancerous feelings, at the age of nineteen, on November 21, 2017, Ellen committed suicide.

I cannot put into words the impact her suicide had on me and many others who loved her. Words like "gut-wrenching," "devastating," "heartbreaking" fall short. Imagine me reading her post, sitting across from another former student in a café, a month too late.

# Step 5: Grieve

The action step for this chapter is grieving.

When I wrote the following lines in my study on grief, *Be A Good Soldier*, I couldn't have known that they anticipated the suffering I would feel at the loss of Ellen. "Putting grief into writing is a clearly rebellious strategy on the part of authors who are, at the same time, aware of their own internal mechanisms that seek to prohibit the circulation of such painful issues."[24]

I ask you to grieve because the bullying paradigm wants you to suppress your sadness. Grieving is an act of rebellion against the bullying paradigm and all the suffering it dismisses, denies, and causes. The more we become aware of the internal protective mechanisms that want us to avoid suffering and struggle, the more we can harness our deep feelings of empathy and compassion for ourselves and others.

I failed to save my son Montgomery from abuse. I failed to save Ellen from bullycide. Both of those failures cause me endless suffering. Grieving helps ease it. I have encountered many who don't want me to write about the bullying paradigm. They find the circulation of such painful issues unbearable. I know how they feel and have empathy for them, but I need to write and rebel. The lives of children are on the line.

## CHAPTER 6

# Return Your Brain to Its Golden Potential

## *Step 6: Train Your Brain*

After Ellen committed suicide, I spent a lot of time sitting on the couch staring at the wall. I was beyond tears. Sometimes my legs would shiver and shake and I'd have to get up and walk around to make it stop. I couldn't really read. I'd just flip through books, sometimes taking a moment to read a paragraph or a sentence. I wasn't able to focus. One unanchored morning, I picked up the work of psychiatrist and professor of psychobiology Jaak Panksepp. The complexity and density of his work, for a nonscientist, made it tough work to understand. My attention was caught by his diagram that lays out our mammalian and essential emotional reactions: fear, seeking, panic, and rage. It struck me as a richer way of understanding the bullying paradigm. Fear is where he locates freezing and fleeing (responses of a victim). Seeking is "appetitive behavior" (conduct of a person safe from bullying and abuse). Rage, or attacking, biting, and fighting, is clearly the realm of the aggressor. The one that I had not learned about was panic. Panksepp notes that it manifests as "motor agitation." The shaking of my legs was panic? He goes on to say it's "isolation induced" and connects to seeking, specifically "seeking social contact."[1] It was an illuminating way to think about what had happened to my brain. On a deep level, I was in a state of panic.

I was isolated in such a profound way that I felt like I no longer belonged to such a bullying world. I felt driven out of society into an "out-group" because I could not go along with the pretend-world that said Ellen killed herself because she suffered from mental illness. That was only half-true. The other half of the truth was that she had been repeatedly abused by adults in positions of trust and power, adults who were in loco parentis. As I closed Panksepp's book, my hand caught on the page where he dedicated his book to his "lost child, Tiina." My legs stilled; tears started down my cheeks. Panksepp followed his dedication to his daughter with a poem by Anesa Miller called "A Road Beyond Loss." It was a poem about being traumatized by grief and slowly recovering through writing. The words resonated for me and opened up a way for me to return.

I no longer felt isolated. Reaching out to my books, learning from and communing with writers was how I could row "on my paper barque" and build a "ladder reaching high above / to light and sound and friends." Like Panksepp, I needed to climb "out of the grief that has no end."

In the late 1990s, American physicians Vincent Felitti and Robert Anda launched an extensive research project to see if there was a relationship or correlation between abusive situations in childhood and midlife chronic diseases like cancer. It is referred to as the Adverse Childhood Experiences study (ACEs). From 1995 to 1997, they undertook a vast study of patients in the Kaiser Permanente health-care system. They surveyed close to twenty thousand patients about adversity in childhood and their present medical conditions. The physicians discovered that traumatizing childhood experiences at the hands of adults—divorce, domestic violence, an alcoholic or drug-addicted family member, or one who was in jail—were indeed precursors to midlife chronic disease. They learned that emotional, sexual, and physical abuse, along with physical *and* emotional neglect, are precursors to midlife chronic disease.[2] Their findings were definitive and undeniable that a history of bullying, abuse, and corresponding high stress not only lays the groundwork for a whole host of mental health issues in midlife, but it also contributes *significantly* to chronic health crises at an older age and thus leads to a shorter life span. You don't just increase your likelihood of depression, anxiety, borderline personality disorder, narcissism, and so forth in midlife when you've been bullied and abused; you also increase your likelihood of cancer, heart disease, diabetes, arthritis, and auto-immune disorders.

Felitti and Anda were startled to learn that childhood adversity and abuse are vastly more common than recognized or acknowledged. While the bullying paradigm constantly tries to suggest there is little to no bullying, or that abuse is exaggerated and designed to harm the reputation of perpetrators, in fact, abuse is rampant and leads to a shortened life span for the majority of victims. Psychologist Alison Gopnik writes that "our job as parents is to provide a protected space of love, safety, and stability in which children of many unpredictable kinds can flourish."[3] The ACE study revealed that for far too many children, love, safety, and stability were usurped by aggression, danger, and instability. Trauma to the brain can even occur in utero. Merzenich points out that "when a woman is living under stressful conditions during pregnancy—from spousal abuse, addiction, poverty etc.—the gene expression in her brain and body are set for a high-threat survival mode." He explains that this translates to the baby's genes so that even before entering the world, the child has "all of the gene-regulation alarms going off in their little brain." Merzenich resorts to capital letters when he feels very strongly about something, and at this moment he stated that one of the most important responsibilities in medicine today, with the knowledge we have, is to "TAKE CARE OF AND MANAGE THE BRAIN HEALTH OF MOTHERS." While this is deeply important for mothers, it is also crucial to allow children to begin life on an even playing field, with golden brain and health potential intact.[4]

Felitti and Anda's research was revolutionary, and it shook the medical community.[5] They measured and documented adversity in childhood that occurred at the hands of adults, *not* the kind of adversity that came from outside events like immigration, natural disasters, war, or being a refugee. Adults raised in a bullying paradigm passed on their adversity to their own children in a cycle that is dismissed and denied but manifests in extremely poor health outcomes. Felitti and Anda documented the way in which abuse in childhood resulted in all kinds of at-risk and self-destructive behaviors that further compounded these children's compromised health and well-being. Many of those who had been abused in childhood developed dangerous aggressive tendencies or a Mind-Bully who turned the aggression inward in the form of eating disorders, addiction, destructive relationships, self-harm, and suicide.

Felitti and Anda's research demonstrated that addiction was not only substance-based; it was "experience-dependent in childhood."[6] In other words, the lure of cocaine or the genetic predisposition to smoking or alcohol were factors in addiction, but they paled in comparison to whether an individual had suffered repeat adversity—at the hands of adults—in childhood. It revolutionized the medical community: "Old assumptions were discarded and a new paradigm established."[7] The new paradigm recognizes that when adults cause adversity to children, it does not stop when they grow up. They might appear whole and healthy, but that's only because we are looking at their bodies and not their brains. The *brain* carries the scars of adversity and abuse into adulthood and impacts the ways in which child-victims treat themselves and others in later life.

The medical community, accustomed to responding to midlife chronic disease as informed by unhealthy choices—such as smoking, inactivity, excessive food consumption, drugs, and alcohol—learned that these unhealthy choices frequently were the result of a traumatized brain. One might assume this realization would have launched an intensive change in how we care for children; educate parents, teachers, coaches, and all frontline workers with kids; put in place protections; and ensure abuse is halted the moment it is discovered or reported. Remarkably and tragically this has *not* occurred. Merzenich, as per usual, goes straight to the heart of the matter: "We modern humans have substantially bought into the notion that human achievements, successes and expressed abilities—and conversely, our disappointments, failures and degraded lives are respectively to our credit, or our fault."[8] The ACE study provides extensive evidence to expose the faulty thinking behind these assessments of ourselves and others. A far more accurate and realistic evaluation must factor in the environmental brain damage individuals may have suffered or their good luck to have had protected brain histories. Deploring others' failures or blaming ourselves mercilessly for our own simply fails to account for science. It would be like faulting someone for suffering from cystic fibrosis or cerebral palsy. Granted, we all have personal accountability and we all must strive to make wise and healthy choices, but when we or our fellow human beings don't, we also need to resist seeing their struggles as being divorced from their childhood experiences, a crucial time of development when they did *not* have control over their decisions, brain health, or overall wellness.

Merzenich stresses this point. "What a poor understanding we modern citizens have about the neurological origins of good or bad, success or failure, capable or unemployable, addicted or virtuous, spirited or mean-spirited, or a dozen other ways we credit and applaud, or find fault and blame our fellow citizens."[9] Twenty years have passed since the ACE study, and we continue to hear about rampant abuse throughout society, so that kids can't be assumed safe in homes, schools, sports, arts, clubs; at the hands of doctors or priests; or any-where. There may have been a paradigm shift in the medical community and the neuroscientific community, but it has not altered in any effective way how we handle bullying and abuse. In fact, the vast majority of people have never even heard of Felitti and Anda's revolutionary study.

The ACE study had sixty-eight questions.[10] (You can take the questionnaire if you would like to see where you land on the ACEs scale, but only if you're feeling strong, have a support network nearby, or are working with a mental health professional. For those of us who have been bullied or abused, these ques-tions can trigger traumatic memories, so please take care of yourself first and foremost.) Each question that focuses on adversity specifies childhood, namely, "before your eighteenth birthday." That is not to say that bullying and abuse later in life are not relevant, but when you are a child, the impact to your Mind-Body-Brain relationship is much more severe because you have little experience, knowledge, escape routes, or independence. In other words, protecting and sav-ing yourself is extremely difficult until later in life.

I think the survey makes a serious error of suggesting that abuse can only happen to children in the home. It keeps asking you about your "household" as opposed to school, church, clubs, classes, or the doctor's office. The shock-ing number of gymnasts, Boy Scouts, students, or Catholic Church victims, if they responded to questions about their household, would not have identified themselves as having been abused. I think it is crucial to widen the concept of "household," if relevant to your childhood experience. The questions seek to understand how often abusive behaviors occurred. This reminds us that "what fires together, wires together." A single trauma might not have the same force to "break" your brain as a repeat trauma. That said, sometimes a single incidence of being shamed, berated, threatened, beaten, or sexually abused can be enough to traumatize someone's brain. We all have unique brains, and all react differently, so you'll have to factor that in to your own responses to bullying and abuse. Here is a pared-down version of the questions that pertained to abuse. I include them as a reminder that Felitti and Anda put emotional abuse on the same plane as physical and sexual. While not specifying brain scans, their research confirmed that harm done exclusively to the brain (emotional, verbal, and psychological abuse) is as serious as harm to the body and brain (sexual and physical abuse). The question about emotional abuse also describes assault, which does not require physical blows for it to be seen as threatening and traumatizing. Here is the emotional abuse question from the ACE study: "Before your eighteenth birthday, did a parent or other adult in the household often or very often . . . swear at you, insult you, put you down, or humiliate you? Or act in a way that made you afraid that you might be physically hurt?"[11]

Here is the physical abuse question from the ACE study: "Before your eighteenth birthday, did a parent or other adult in the household often or very often . . . push, grab, slap, or throw something at you? Or ever hit you so hard that you had marks or were injured?" Here is the sexual abuse question from the ACE study: "Before your eighteenth birthday, did an adult or person at least five years older than you ever . . . touch or fondle you or have you touch their body in a sexual way? Or attempt or actually have oral, anal, or vaginal intercourse with you?" The sex abuse question *should* include a question about grooming along the lines of: "Before your eighteenth birthday, did an adult or a person at least five years older than you ever . . . offer unwanted affection and attention, massages, hugs, privileges, invitations to events, gifts, money, sexual jokes, tickling, comments about your appearance like how beautiful or sexy you are, send inappropriate texts and tell you to delete them?"

Here is the emotional neglect question from the ACE study: "Before your eighteenth birthday, did you often or very often feel that . . . no one in your family loved you or thought you were important or special? Or your family didn't look out for each other, feel close to each other, or support each other?" Here is the physical neglect question from the ACE study: "Before your eighteenth birthday, did you often or very often feel that . . . you didn't have enough to eat, had to wear dirty clothes, and had no one to protect you? Or your parents were too drunk or high to take care of you or take you to the doctor if you needed it?" As noted, the other five questions on adversity in the ACE study are related to family crisis such as divorce, addiction, incarceration, mental illness, and domestic violence. Beyond the ten questions about childhood adversity, the other fifty-eight ask the patient about themselves, their behavior, and their historical and present health.

When you answered the ten questions about adversity in childhood, if you scored zero on the ACEs scale, the researchers found that you would be much less likely to suffer depression, become obese, develop addictions, have anxiety, or commit self-harm or suicide. In contrast, if you had a score of four or higher on the ACEs scale, you would be much more likely to suffer many of the above chronic conditions. To put the data in sharp perspective, you have an 80 percent greater chance of attempting suicide when your ACEs score is four or above. Now if you took the survey and scored at four or above, this might be very painful information to hear. Once again, it is important to ensure that you have a network of supporters to reach out to and a mental health practitioner who can care for you. This information can be upsetting and can trigger a reaction, thus caring for yourself is paramount. That said, ignoring this information can act for some of us as a major block to actually setting in motion a healing program. If we don't know our brains were hurt in childhood, and the invisible neurological scars we have continue to shape our present-day behavior, we are unlikely to turn to neuroscientific research for ideas on how to heal the scars and restore our health.

What's significant about the ACE study is that the almost twenty thousand Kaiser Permanente patients who were surveyed did *not* know that the abuse done to them as children was shaping their health outcomes and their potential at-risk behavior as adults. Considering that these risky behaviors put their health at risk and shortened their life span, surely it is critically important knowledge to have. What's even more significant is that most doctors did *not* know either. Felitti and Anda themselves were shocked by their findings in the ACE study. Who knew that so many children suffered terrible adversity at the hands of the adults in their household? Who knew that the bullying and abuse done to many children would result in severely compromised health and a potential series of health-risk behaviors, that then made a ton of middle-aged people sick with chronic disease and slated for earlier deaths? As noted, those who abuse are rarely held accountable; thus, the health issues appear to be *your* own fault because *you* don't take care of yourself. This is why it's so important for us to keep distinguishing between our mind and our Mind-Bully. It's when you internalize, normalize, and listen to the bullying voices within that you mistakenly take responsibility for what was done to you rather than, with insight and compassion, set about healing your Mind-Brain-Body.

What was the catalyst that set Felitti down the path to issue the ACEs questionnaire to almost twenty thousand patients? He was treating a female patient in later life for coronary artery disease, and Felitti asked her to tell him about her life history. She traced her morbid obesity back to her chronic depression and then back to childhood sexual abuse. The patient's declining health trajectory concerned Felitti, and he started asking other obese patients about their lives and discovered that the majority of them had been sexually abused as children. This discovery launched his research into the relationship between childhood adverse experiences and chronic disease in midlife. The children who were frequently belittled and berated, threatened or ignored, the children who were spanked or beaten, the children who were sexually groomed and abused by adults, the vast majority did not have visible injuries or scars in midlife. The damage was invisible. It was inside their skulls. The harm to their brains was correlated with the ways in which they became self-destructive or aggressive. They developed eating disorders, resorted to self-harm, suffered depression and anxiety, self-medicated with alcohol and drugs, pursued unhealthy relationships, engaged in violent interactions, and committed suicide.

Felitti stepped out of the old paradigm that treated disease symptoms as resulting from present-day behavior and the body and ignored harm that might have occurred to the patient's brain during childhood. Felitti and Anda entered into a new paradigm that questioned patients from a holistic point of view. Ordinarily, medical approaches revolve around the body while ignoring the brain altogether. Merzenich notes that doctors throughout your lifetime check your pulse and your blood pressure, do blood and urine samples, do EKGs to ensure your heart is healthy. Then he asks a worrisome set of questions as to when your doctor last checked you for "forebrain connectivity," examined your "continuous brain activities that support complex operations like serial memory and thought," or measured your "processing speed." For Merzenich, these are rhetorical questions because he already knows the distressing answer, which is "never."[12] How is it possible that experts care for our bodies but they do not offer the same care to our brains? This is reflected in how our laws protect bodies but fail to protect brains in the same way. If the law can see harm done to the body, it holds the abusive individual to account. If the law cannot see the harm done to the brain, it rarely holds the abusive individual to account. Likewise, medicine sets out to heal a body injured by bullying and abuse, but it does not regularly (or ever) heal brains injured from bullying and abuse.

The bullying paradigm is so entrenched that three decades of neuroscientific research have not stopped us from ignoring our brains until they are in irreparable crisis. Merzenich explains that a brain needs to become a "train wreck" before medical experts will act on its behalf, and by then, it's too late. It's like we dismiss and ostracize our own brains and it's normalized in our society. The bullying paradigm leads us to be "caricatures" of what we could become. We seem unable to demand more than an indoctrinated brain, limited by a training framework that acts as though the height of our brainpower is to obey. We squander our own golden potential due to suffering childhood adversity. We ruin our own health because of what has been done to our brains in childhood. The adults that do the damage are rarely, if ever, held accountable, let alone rehabilitated so that they do not simply pass on their own adversity to the next generation. This is the infection, the epidemic that we need to recognize so we can set in motion a brain-health plan. First of all, we need to shelve the outdated moral terminology that we apply to unhealthy behavior and instead use medical terminology.

In an extraordinary discovery, Felitti realized that obesity was not a "sin." It wasn't about a flawed character who indulged in "gluttony" and hence was an exemplar of moral failing. The bullying paradigm is quick to shame someone who is obese and blame them for their condition. Felitti realized those judgments were outdated and inaccurate. In the majority of cases, obesity was a conscious or unconscious choice on the part of abused patients to keep themselves "safe." A woman who had been raped in childhood wanted to send a message to others who may attempt to violate her. A man who lived and worked in a violent neighborhood wanted to appear huge to deter attacks. Despite "social taboos against obtaining this information," red flags of the bullying paradigm, Felitti made a courageous decision to document childhood abuse and see if it had any correlation with his patients' health issues in later life.

While Felitti and Anda were replacing the outdated model of medicine that saw the brain and body as unrelated, in the realm of education, neurologist Antonio Damasio was undertaking research that supported their findings. In 1995, Damasio questioned the erroneous division we have operated under since seventeenth-century philosopher René Descartes announced, "I think, therefore I am." Descartes launched a highly influential paradigm that focused on our cognitive function, our reason, our rational minds, all the while dismissing our emotional function, our feelings, our embodied intelligence. Damasio's recognition of the anomaly, what he calls "Descartes' Error," has allowed him to work within a new paradigm whereby the brain is not confined to the world of thinking. The brain is recognized as the integral, connected, inseparable part of ourselves that depends on the feelings of the body to make rational decisions. Damasio discovered that patients who have brain injuries that block access to their emotions cannot make rational decisions. He recognizes that a separation from brain and body is false and artificial.[13]

After receiving overwhelming confirmation that childhood abuse and midlife chronic disease went hand in hand, Felitti's response was to issue a call to action. He identified the false mind/body split. He exposed the flaw in the outdated scientific system where doctors assess patients as physical bodies in crisis and prescribe surgery, pharmaceuticals, or other medical interventions. He points out that this way of diagnosing patients is "well protected by social convention and taboo," as we do not want to know the origin of the body's sickness, the patient's suffering, the abuse history that led them there. Felitti explains that doctors approaching patients without asking about abuse and trauma have "limited [them] selves to the smallest part of the problem." They can safely and comfortably remain in a silo where their job is to administer medication that treats the body while ignoring the brain. This is what Helen Reiss refers to as "downstream medicine," which treats illnesses rather than seeking their origins and striving to prevent them. Felitti voices his concerns in identical terms, arguing that the medical community treats "tertiary consequences far downstream" instead of preventing one of the most significant sources, namely childhood abuse.[14]

In the medieval era, educated individuals believed in a miraculous transformation through alchemy. They believed that they could transform matter from regular or base metals into gold. Felitti asks a tragic question: "How does one perform reverse alchemy, going from a normal newborn with almost unlimited potential to a diseased, depressed adult? How does one turn gold into lead?" Once we become aware that our brains may have suffered neurological scars from childhood abuse, then we can shift our focus to healing and restoring our health. We need to reverse what Felitti sees as "reverse alchemy." If parts of our brains have been turned to lead due to childhood adversity, then who is to say we cannot return the base metal back to its original golden form?

Felitti wonders: "What do these findings mean for medical practice and for society?" He summarizes the significance of his and Anda's massive research project: "Clearly, we have shown that adverse childhood experiences are both common and destructive. This combination makes them one of the most important, if not the most important, determinants of the health and well-being of the nation." What's empowering about his summary is that if childhood adversity is the most important determinant of our health and well-being, well then, it's up to us to heal. If harm can be done to our brains so that our Mind-Brain-Body is fragmented and works at cross-purposes, well then, surely we can turn to our cerebral command center and change that. The experience of childhood abuse is "common." While victims often feel as though they are outliers, full of shame that blocks them from truly belonging to the community, it's not true. It's yet another myth propping up the bullying paradigm. If you've been abused in childhood, you are in the majority. You are part of a vast network of suffering individuals who can work together to repair the harm done to their brains.

Anda calculates that the overall costs stemming from abused children exceeded those of cancer or heart disease and that "eradicating child abuse in America would reduce the overall rate of depression by more than half, alcoholism by two-thirds, and suicide, I.V. drug use, and domestic violence by three-quarters. It would also have a dramatic effect on workplace performance and vastly decrease the need for incarceration."[15] We pour resources, we task brilliant minds, we have vast organizations devoted to curing cancer and educating us about heart disease, but we fail to eradicate abuse in childhood. The pressing question is: can we reverse the "reverse alchemy"? Can we return the brain to its golden health and potential after childhood adversity has turned it to lead? In discussion with Merzenich, I found out that the answer was simply *yes*. *Yes*, the brain can be retrained back to wholeness, health, and golden potential.[16] Isn't it tragic that Ellen realized on a cellular level that the impact of childhood abuse had put her at risk from "cancer" but that she had not known there was a cure?

When I was reading Merzenich's unpublished manuscript on how to prevent various traumas to the brain, a section jumped out at me because it sounded similar to how a brain could become disordered like Ellen's had. Merzenich explains that in the labs of neuroscientists, it was possible to catastrophically dismantle the brain in specific ways. Sounding more like a terrifying sci-fi novel, scientists are now well-equipped with their knowledge of neuroplasticity to train your brain in such a way as to "destroy the functional use of your right leg," "turn your hand into a useless claw," or "destroy your ability to understand what I am saying, or writing, or thinking." The line that really impacted me was that the scientists could use targeted training to "drive you into a deep and (if you like) suicidal depressive swamp."[17]

As Ellen said, she felt trapped in such a place and no one in her world—including medical professionals—seemed to have a ladder to help her ascend from the darkness of the "depressive swamp." The ladder is exactly what has motivated decades of research done by Merzenich. It echoes the poem from Panksepp's dedication page, the "ladder reaching high above / to light and sound and friends." The question that Merzenich sought to answer was: "How many of these different physical and functional indices of brain function and organic brain health can be overcome via intensive, progressive brain training?" And the answer has been a resounding "ALL were recovered, by relatively simple forms of training."[18] Merzenich's forty years of research into neuroplasticity has revealed that while brains can become terribly harmed, injured, and disordered, neuroscientists and the insights they've developed *can* retrain them back to health. They can even take traumatized brains and make them *stronger* than they were before.

In initial discussions, I was struck by how humble and affable Merzenich is. He refers to himself as "Mike" and doesn't stand on ceremony. While he is undeniably sharp, he softens his mental acuity with clear curiosity and interest. Because the stakes were so high for me on an emotional level, I was extremely nervous when I first met with him. I had planned out and rehearsed what I needed to say so as not to waste his precious time. I knew he received more than a hundred e-mails a day, despite being formally retired. The world rarely lets geniuses rest regardless of how much they may have earned it. After I explained my profound worries about child abuse, the broken system that enabled rather than halted it, and my realization that neuroscientific research was the antidote, he smiled and said: "How can I help you?" It made my eyes sting with tears because for years I had been coming up against the wall of the bullying paradigm that was constructed by those who preferred or benefited from the status quo, regardless of how many kids were suffering, self-destructing, or even killing themselves. Merzenich's question gave me a jolt of hope that I was on the right track, that indeed, neuroscience has insights that can be applied to heal neurological scars caused by bullying and abuse and thereby restore the health of victims and perpetrators (who were likely once victims).

After he generously shared with me his various works-in-progress that had not yet been published, we had a meeting. I quoted from the section of his manuscript on the various ways that neuroscientists could "drive organic brain health in a sharply negative direction, driving it at high speed to get 'very old' almost overnight." I asked him if this enactment of advanced, rapid aging, in an otherwise healthy brain, could be compared to what abuse did to the brain? I wanted to know if it was crazy of me to think that Ellen had suffered a very sped-up dementia that resembled the kind of brain malfunction an elderly person might suffer if they had Alzheimer's. Merzenich confirmed that this was indeed a fair comparison.[19]

As I shared with you before (but it bears repeating at this juncture), Merzenich's research has shown that the "simplest way to accelerate brain aging is to simply continuously bombard the brain with neurological noise—continuous 'chatter' that frustrates the brain's ability to record or recognize 'what's happening.'" Now, after hearing directly from a traumatized brain like Ellen's, imagine what happens to a brain that is bombarded by the grooming, the abuse, the gaslighting, the hypocrisy, the Orwellian doublespeak, the outright lies, and the cover-up that form the foundation of the bullying paradigm. As Merzenich writes: "When the brain cannot get the answer right, which it cannot under these conditions, it rapidly changes ALL of its operational characteristics in a degrading direction. *Everything goes to hell together*."[20] Ellen's description of what was happening in her brain, which had begun to threaten and attack her body, had finally found a scientific articulation. Her post on Facebook revealed an utterly brilliant brain struggling against the bombardment of the bullying paradigm.

Wanting to be very clear, I asked Merzenich, with a sinking heart, if in fact this bombardment to the brain was comparable to what abuse did. "Absolutely," he replied.[21] While I knew he had dedicated his life to saving people from brain traumas of all kinds, I was struck by the apparent emotion his answer conveyed. Merzenich continues to be astounded that the outdated paradigm ignores the brain, despite thirty years of scientific breakthroughs of epic proportions. What is crucial for us to understand is that his response *absolutely* goes both ways in neuroplasticity. While the brain can be changed and degraded by abuse, it can also *absolutely* be retrained back to health. Merzenich did not earn the Kavli

Prize for Neuroscience in 2016 because of his research into brains that are falling apart; it was because he has learned how to put brains back together.

Merzenich told me that it's crucial for readers to understand that the ACE study is yet more confirmation that brain and body health are causally linked. "The brain plays a key role in the autonomic nervous system and hormonal regulation. Those processes contribute, in turn, to immune response powers and regulation." Put more simply, your brain governs the health of your body due to their inextricable influence upon one another. When brains are healthy, they maintain homeostasis or optimal functioning and balance in your body organs, as well as respond to innumerable challenges, thereby contributing in important ways to "immune system powers." In other words, an unhealthy brain, a damaged brain is going to struggle to keep the body healthy. As Merzenich confirms, "It is no accident that long-lived people usually sustain healthy brains right up to the end of their longer lives."[22]

How do we become one of those people who have a healthy brain right up until the end of our lives? Any one of us can go online and start exercising our brains using the Merzenich method, which has been documented in more than a hundred peer-reviewed studies to make significant improvements to brain health at any stage of life. We can prevent the onset of age-based brain atrophy, and we can heal the scars of various brain inflictions, including trauma resulting from bullying and abuse. In other words, Merzenich has designed inexpensive, easily accessible, online programs that silence "brain noise," which he describes as "poison."[23] At minimal cost, the programs are accessible online at www.brainhq.com. Doing these brain exercises or workouts has been proven to help a brain emerge from the "depressive swamp of suicide," as well as ward off a much slower brain decline that occurs naturally as we age and ultimately for some of us may present as Alzheimer's. The question that haunts me is: why are so few of us informed about these neuroplasticity programs? How can we help an abuse victim like Ellen when we are so ignorant about our brains and the remedies designed by neuroscientists? An even more concerning question is: Why don't more mental health professionals, like the ones assigned to Ellen's care, seem to know about the evidence-based programs designed by neuroscientists to heal traumatized brains? Merzenich's four decades of groundbreaking research that provides evidence for our "soft-wired" brain were fueled by his "soft," empathic heart. He did not read about Ellen's suicide with a rational mind, safely ensconced in a research lab. He wrote to me: "How cruel can it be, to understand the true origins of so much of the struggling amongst us, **knowing that those struggling individuals can be helped**, yet stand by and do nothing TO help them?"[24] The emphasis, conveying his voice, belongs to Merzenich.

My goal is not to blame and shame. It's to point out just how trapped we are as a society, as a medical-specialist and mental-health-specialist community even, in an outdated paradigm that is more likely to prescribe pharmaceuticals than to harness neuroscientific breakthroughs. While people are pouring billions into trying to find a pill to offset Alzheimer's, a prevention of its onset already exists in Merzenich's brain exercises. While suicide is the first or second leading cause of death in youth populations in many developed countries, the vast majority of parents and even professionals don't know there's an antidote to the poison in Merzenich's brain exercises. Over the past two years, Merzenich and his team at Posit Science have been working hard in North America and in Europe to provide access to an inexpensive and relatively easily deployed form of help as fast as humanly possible. They are working in the United States with a "train the trainer" model, running trials with Australian colleagues in Great Britain, translating the program into different languages, and striving to harness technology so that they can offer rapid implementation of brain-training "help" across the planet. When Merzenich said to me, "How can I help you?" I did not realize that this is the question he is directing to every nation and every individual in the world. And I did not realize until now that he already has the answer. He already knows exactly how he can help heal our brain scars and restore our health.

Merzenich has adapted his programs to create a version for young children called "Stronger Brains." He codirects a not-for-profit with Wendy Haigh, who is based in Australia and who has dedicated her life to helping *all* children, not just those whose families can afford intervention (www.strongerbrains.org). Haigh has constructed a holistic program that can support the most traumatized children through brain training back to health. Former president of the British Medical Association and first National Clinical Director for Children in government Sir Al Ansley-Green wrote in 2019: "New work by Merzenich and Haigh is showing that disordered brain 'wiring' can be reversed through new ways of stimulating the brain, and I see this to be one of the most important and exciting opportunities in neurodevelopment today."[25] Merzenich's unfaltering commitment to see the knowledge of neuroplasticity applied to enhance all lives matches powerfully with Haigh's drive to see all children receive affordable interventions to optimize their brain performance, especially when it has been put at risk through trauma.

I asked Haigh to walk me through the program that traumatized children could use to heal the inexpressible and invisible scars on their brain. It was a relief to speak with her; I felt like I'd found a kindred spirit despite the fact that we live on opposite sides of the planet. Haigh's response to the findings of the ACE study is to go straight into action. "We get kids with all kinds of labels on them. We take *all* kids. All brains. If that child has a brain, we should be able to strengthen it. Like we could strengthen their leg. We operate with a policy of inclusion." Listening to Haigh talk about the Stronger Brains approach is so inspiring because it is evidence-based. It works, and that has been extensively documented. Her clarity is stunning. It's really so simple and yet the bullying paradigm would have you believe that stopping rampant child abuse is extremely complex and very unlikely. Doesn't seem like it when you're listening to Haigh. Everything she says applies to adults as well, except that we are far better equipped to do the brain training ourselves. Haigh acknowledges that "the problem is massive and the outcomes for children are getting worse." In fact, when Stronger Brains does whole-of-school brain health assessments, they discover that approximately half the children need brain training. "For example we see 1 in 2 children with low working memory which makes subjects like maths so much harder to do." Not overwhelmed by the scale of the challenge, Haigh's eyes light up: "The good news is we can measurably heal the brain and what's really rewarding is that we see the biggest improvements in the kids that need it most, such as the ones with the lowest baselines or greatest trauma."

# Step 6: Train Your Brain

Haigh clarifies that traumatized kids cannot simply do the online programs at Stronger Brains in the same way that adults can independently work on Merzenich's BrainHQ site. She and her colleagues have learned that the kids' trauma all too often blocks them from being able to focus and settle enough to begin the brain exercises. The program "doesn't work, unless you apply it properly. It's not a quick fix. There's a process involved. When you're injured in your body, you need professional help, same with the brain. These are injured kids." Any adult who externalizes their trauma from abuse, putting their relationships, reputation, and career at risk, is just like these injured kids. Any adult who internalizes their trauma from abuse and puts their health, body, and brain at risk (think substance abuse, self-harm, eating disorders, suicide) is just like these injured kids. What Haigh and her team have designed at Stronger Brains doesn't just apply to children. It is something traumatized adults can also put in place to recover from adversity in childhood.

At Stronger Brains, they don't just look at children's health and well-being by assessing their bodies; in a revolutionary way, they also factor in their *brains*. Even though the injuries to the brain are invisible to the naked eye, Stronger Brains opts to foreground them. Children need a great deal of support before they even begin working on the Stronger Brains program, and I think many traumatized adults do too. As Haigh outlines: "The traumatized kids we work with are trapped and going in circles. Stronger Brains helps them build a ladder out of the well of the emotional brain into the rational brain. We put them on a journey to a stronger, brighter place. Right now they are in survival mode."

Stronger Brains uses a train-the-trainer model as they've found it's more effective to have the kids work with adults with whom they're familiar. Haigh describes the impact of these mentors: "Makes it faster and better for kids. The adult acts as a mentor who motivates, recognizes the child's work, rewards effort and achievement." She emphasizes that with these traumatized kids, they've "been trying their hardest and yet they're constantly lambasted for being defiant, distracted, acting out etc." You can quickly recognize the labels of the bullying paradigm. As Haigh says, "Kids can't just do Stronger Brains on their own, especially when they've been traumatized. They are so easily distracted. After trauma, they can't focus. So we involve adult mentors each step of the way." I know lots of people who have worked with personal trainers who want a healthy, strong, flexible set of muscles, but I have *never* heard of someone hiring a "brain trainer," even though it is just as important if you want a healthy, strong, flexible set of neural networks.

Haigh stresses that in the train-the-trainer program, Stronger Brains ensures that the mentors—whether they're parents or teachers—learn about how the executive function of these children's brains is so traumatized that they struggle to process and that will slow them down. "These kids can't process fast enough. The mentor needs to be informed, trained to understand *why* the kids perform this way. Parents and teachers need empathy, patience and understanding for the kids as it's hard work and a slow process to build new neural pathways to restore the brain to neurological normalcy." Restoring the brain to health generally takes less than one year. With thirty minutes of practice, done daily, Stronger Brains repeatedly sees lasting transformation.

Once the kids are calm and motivated, then the actual brain training begins. Depending on how severe the trauma is, sometimes all a child can handle is five minutes of focus. Over time, however, the children slowly but surely improve their concentration and commitment until they are clocking in thirty minutes a day. Remember that if you begin and find that all you can bear at the start is five minutes, then that's understandable and praiseworthy. Give yourself time to slowly build up your neural strength. On Merzenich's BrainHQ site, you can set in motion your own "Personal Neuroplasticity Plan" and know that the program will meet you where you are. Some parts of your brain may already be very strong, while others might need training.

At Stronger Brains, they slowly but surely move the children from survival mode back into brain health and happiness. They begin with "Neuroeducation" to teach a brain-health curriculum designed by Merzenich. The key message is that each child has a unique brain full of potential, and they can make their brains stronger to become a "better captain of their ship." Haigh says that the children have a "light bulb moment" when they realize that they have the power to strengthen their "CEO" or prefrontal cortex, their captain, by putting in effort. It is in this moment that children realize they're not "hopeless" or helpless and that Stronger Brains has a solution to their struggle and a proven track record in strengthening brains.

Once the kids believe in themselves, the next stage is to calm their brains (Personalized Empowerment Plan). Then they can focus on the transformative part of the program, the brain training, whereby they build new neural pathways (Personal Neuroplasticity Plan). Stronger Brains evaluates the neurological weakness of each child or youth by applying a battery of scientifically validated assessments developed by Merzenich and his team of neuroscientists. The program then targets any brain weaknesses with "intensive, adaptive, gamified" training. This brain training can overcome "even long-standing neurological impairments." Haigh and her team "continuously track performance improvements to ensure that the targeted neurological changes are achieved and neurological normalcy is restored." Haigh shared her and Merzenich's dream:

> Our vision for the future is that all kids in all schools will receive regular brain assessments and the practical help and healing needed to thrive, just as we expect if they injure their leg. One of our schools is already delivering our program on a whole-of-school basis and integrating our services with others like health services onsite for both parents and students.

This neuroscientific approach sounds more like being in a doctor's office or a scientist's lab than in some outdated schoolhouse where the bullying paradigm tries to convince us we're involved in a knock-down-drawn-out battle about "character." A science-informed approach leads us out of the house of religion where the bullying paradigm tries to convince us we're involved in a contest for the "soul." Don't get me wrong, characters and souls are worth fighting for, but when it comes to assessing an individual's health and well-being, it's also important to include the language of science. Haigh assures me that Stronger Brains is a joyful place for students. She told me about one boy who began the program highly depressed and withdrawn. He had dropped out of school and was not working. This boy traveled two hours to participate in Stronger Brains and then two hours to get home every day. He'd arrive "early each morning to cook breakfast for his classmates who may not have eaten. He'd stay in the afternoons, volunteering his help with interviews or anything that needed to be done." Haigh's face lit up: "He was a delight."

Stronger Brains has been successfully delivered in schools, colleges, and prisons. It's been successfully delivered to employment, disability, and foster services. Haigh shares with me that Stronger Brains has been evaluated by the Australian government through the University of Queensland and University of Melbourne as part of their "Try, Test & Learn" initiative, which was a pioneering study that looked at innovative ways to reduce welfare dependency for young people at high risk of long-term unemployment. She explains that as "with many countries, the welfare budget is the largest and fastest growing area of government expenditure and not sustainable, so the government was seeking some game-changers. Stronger Brains was included as one such possible game-changer with the potential, if successful, to scale up nationally." Haigh and her team ran a program called "Rewire the Brain," which showed "statistically significant outcomes" achieved by hundreds of youth in education, mental health, and employment who had been identified as high risk for long-term unemployment. You can see why Haigh and Merzenich see their work with Stronger Brains as the number one moral, social, and financial imperative of our time.[26]

Wouldn't Drs. Felitti and Anda be thrilled to see Merzenich and Haigh's neuro-community that is building up around Stronger Brains? While the ACE study revealed just how significant the harm is to traumatized brains from abuse, Stronger Brains reveals that "reverse alchemy" can itself be reversed. Brains *can* heal their scars and restore their golden health potential. It's an uphill battle due to our brains' "negativity bias," which imprints adverse experiences much more deeply and indelibly than good ones. Neuroscientist Rick Hanson captures this hard truth with the idea that our brain is like "Velcro for bad experiences." They really stick. At the same time, our brains are like "Teflon" for good experiences.[27] They just slide away. Like Stronger Brains, Hanson is not daunted by this neurological fact. His research into ways that we can hardwire happiness into our brains shows that with "*safety, satisfaction, and connection*"—the three pillars that uphold the child-saving work of Haigh and Merzenich—we can heal.[28] What sets Stronger Brains apart is that these pillars uphold scientifically validated brain training. In a world that has not yet discovered the healing power of brain training, in any kind of widespread way, I felt an incredible surge of hope and a desire to share this crucial information with all victims, all teachers, all parents, and especially all children. The alchemy is real. Your golden potential is just waiting for you to work with your brain.

That said, after speaking with Haigh and learning in detail about the transformative power of Stronger Brains, I am sorry to report that I was sabotaged by my Mind-Bully. I came face-to-face with the painful truth that I really didn't believe in myself. I didn't think I could achieve the goal of strengthening my brain. There was something blocking me. There was something wrong with me. Although I knew it was going to be painful, I had to find the courage and resolve to return to the repeated abuse I suffered as a teenager at the hands of three teachers. Like my student Ellen, I was sexually harassed, groomed, and pressured. Like my son Montgomery, I became the target of relentless humiliation as I resisted the teachers' sexual harassment and rejected their sexual advances. I needed to face that dragon first if I was going to develop a stronger, healthier, happier brain.

# CHAPTER 7

# Stop Identifying with the Abuser

## *Step 7: Believe in Yourself*

In 1973, escaped convict Jan-Erik Olsson walked into the Kreditbank in Stockholm, Sweden, carrying a loaded submachine gun. He immediately fired a round at the ceiling, sending down a "shower of concrete and glass."[1] Then he tied up bank employee Birgitta Lundblad and secured the wrists and ankles of teller Kristin Ehnmark. Olsson ordered another employee to tie up cashier Elisabeth Oldgren. Everyone else exited the bank while Olsson demanded the police send for another convict, Clark Olafsson, with whom he had done time at Kalmar penitentiary. A police negotiator suggested that Olsson exchange the hostages for the company of Clark. In response, Olsson "grabbed Elisabeth by the throat and jammed the submachine gun into her ribs."

The summoned convict, Clark arrived in the afternoon, and later on, while touring the bank, he discovered a junior bank employee hiding in a stockroom. Sven Säfström became the fourth hostage. Olsson decided to hole up with Clark and the hostages in a vault that had a "close, oppressive quality." Elisabeth told him she was getting claustrophobic. Putting a "length of rope, perhaps thirty feet, around her neck," he let her out for a walk. Elisabeth reported after, "I couldn't go far and I was on a leash that he held, but I felt free. I remember thinking he was very kind to allow me to leave the vault." A woman, with a rope around her neck, feels "free." A woman, permitted to exit a suffocating room in which she's trapped with others, is struck by her captor being "very kind." A strange reversal is occurring.

Elisabeth was not the only captive to be impressed by Olsson's kindness. Nor was she the only one to be leashed by a rope and infused with a sense of freedom. When Kristin was permitted to leave the vault and go to the bathroom, she saw police hidden within striking distance. "One of the officers asked Kristin in a whisper how many hostages the robber had taken. 'I showed them with my fingers,' Kristin said. 'I felt like a traitor. I didn't know why.'" Again, note the reversal. The hostage feels as though she is *betraying* her captor by providing the police with information. In the cold of night, Elisabeth finds herself chilled and Olsson puts his coat over her shoulders. In an interview, Elisabeth describes Olsson "as a mixture of brutality and tenderness. I had known him only a day when I felt his coat around me, but I was sure he had been that way all his life." The captive Elisabeth is describing the personality split of the classic abuser, a Dr. Jekyll and Mr. Hyde.

As the days passed, the police commissioner requested to see the hostages to know if they were indeed still alive and unharmed. Olsson agreed, and the commissioner was baffled to see that the hostages "showed hostility" *not* to their captors but to him. With their captors, they appeared "relaxed" and "convivial" and even displayed an "easy camaraderie." The commissioner reported to his aides that he was "up against a mystery whose clues, if any existed, were like none to which he had ever been exposed." And then "a wave of theorizing about the captives' odd pliancy" unfolded. Is this really a mystery? Would you describe the behavior of the captives as "odd"?

In contrast, it makes perfect sense that the brain instantly recognized, when Olsson shot off the first round with his submachine gun, that survival meant aligning with power. The more unpredictable, controlling, threatening, aggressive, and dangerous an individual, the more likely people will bow down to their power. The Kreditbank story is an extreme example, but look at any bullying scenario and you'll see brains trying to figure out just how much they might be in jeopardy and aligning themselves with the safest bet, even if that requires identifying with the aggressor. Trapped in the Kreditbank, the captives' mirror system would have gone into hyperdrive so that every single feeling, thought, and intention of the terrifying captor would have been of primary concern for the hostages' brains. When Olsson first set off his submachine gun, Birgitta explains that for her, all that mattered "was his next move." The police commissioner's inspection, if anything, put the hostages at risk when ruled over by an unpredictable and potentially deadly captor.

Listen to the way in which the hostages describe their conflicted feelings in interviews after their release.

> The robber [Olsson], chewing his caffeine tablets, behaved more mercurially than he had before—his moods, as Elisabeth saw them, alternating between a tenderness and a brutality that had both become heightened. Kristin told me that, with the shutting of the door, her fear of death seemed to have sharpened. "I hadn't thought that was possible. I had never been close to death before," she said. "From

> the moment Jan made me his hostage, I was afraid he would suddenly kill me, but now it was the police I was afraid of—even more so than when I had talked with the Prime Minister. I felt hopeless. What difference did it make, I asked myself, which one of them did away with me?"

The convicts and hostages, who spent six days holed up in the Kreditbank together, are at the origin of the psychological phenomenon henceforth referred to as "Stockholm Syndrome." The phrase was coined by criminologist and psychiatrist Nils Bejerot, who was on the scene during the six days of negotiation as Olsson and Clark held the hostages. Bejerot sought to explain how it was possible that hostages could become sympathetic with, and even bond with, their captors. In the aftermath of the hostages' escape and the convicts' return to prison, Daniel Lang, writer for *The New Yorker*, went to Stockholm and interviewed the involved parties. His article was published the following year on November 18, 1974. Lang learned from the psychiatrists involved that the hostages' bonding with their captor was hardly surprising. This mechanism gets activated when people are in "survival situations." Lang reports how the psychiatrists told him that psychologist "Dr. Anna Freud has called the reaction 'identification with the aggressor'" and she traced it back to the "deepest layers of one's being, stirring unconscious memories of one's earliest patterns of security and order." While psychology uses terminology like "stirring unconscious memories," neuroscience documents that the brain is wired to survive by evolution. In a survival situation, the brain is not seeking patterns of security and order. It's making the kinds of instinctual plans you form when faced with a deadly predator. You can well imagine the hostages' hypervigilance, the adrenaline and cortisol pumping through their brains and bodies, the activation of their sympathetic nervous system frantically trying to figure out whether to fight, flee, or freeze. Clearly in a hostage scenario, when your captors have guns, freeze is the best option, and that's exactly what they did. They remained in place. They created a community built upon a shared primal neurological impulse to survive at all costs. This is not mysterious or odd. This is sensible and normal.

Psychiatrist Lennart Ljungberg, who had come from Germany to Sweden in the aftermath of the Second World War and worked with the hostages after their release, explained that to survive, even "Auschwitz inmates had tried to like their captors, just as Sven and the three young women had" bonded with the convicts. The other psychiatrist who cared for the hostages, Waltraut Bergman, explains further that if the hostages broke with their captors, "they might have been overwhelmed by 'fear of death, chaos, and the elimination of all normal laws.'" The law that kept captors and hostages in line, connected, bonded even, was the law of survival. Ljungberg comments: "Each of them very much wanted to go on living," and that included the bank robber.

Kristin described in interviews with police that not only did she consciously bond with the captors as a bid for survival, but also she sought out "tenderness" from them. She would hold hands with Clark from time to time. She told the police that it "made me feel enormously secure. It was what I needed." If someone has immense power over you, it's not a bad idea to connect with them. Child abuse victims who depend on adult care, affection, authority, credentials, assessment, and so on are easily trapped by abusive adults who leverage their dependency to morph the relationship into an abusive one. "Each show of friendliness on Olsson's part reinforced his leadership. As Sven put it, 'When he treated us well, we could think of him as an emergency God.'" At one point, the hostages and captors were desperately hungry, and Olsson revealed that he had three pears, which he shared among them. This deeply parental gesture of providing nourishment for "children" obviously reinforced his power over them and made them want to please him. Abusive individuals frequently swing from a providing, loving parent to an unpredictable, harmful tyrant. The fear that surrounds the "god-like" abuser makes anyone who might break the bond seem like an "enemy" or traitor, as Kristin explains. When children are asked to speak up and report abuse, or even when adults are asked to report domestic abuse or colleagues are asked to report on toxic leadership, they struggle and oftentimes cannot go through with it. A reversal born of fear and a deep-seated will to survive occurs. The "enemy" becomes the one trying to stop the abusive individual: the police, the rule of law, the courtroom, the media. The "enemy" can even become loving parents who are trying to wrest their child back from a cult or save them from a single abusive individual.

As Sven articulates the psychological reversal, it helps explain the entrenched pattern of victim blaming. Here is how he describes what it was like to be a hostage: "We were all sympathy, taking in everything they told us. We acted as though they were our victims, not the other way around." The victim identifies with the aggressor; the abuser becomes the victim; the abuser is the subject of a witch hunt; the abuser denies accountability because he or she is always an innocent victim and those whom they abuse become their staunchest defenders. It's mind-bending reversal upon reversal.

As the hostage crisis in the bank went on, the police became increasingly worried, and they prepared to send knock-out gas into the vault to drive the captors out with their hostages.

> Sven grabbed for the phone. His voice urgent, he shouted, "Don't send in gas, whatever you do." Police lowered, and quickly raised, a camera to ascertain for themselves what was going on. The hostages were in nooses. Olsson, assisted by Clark, had placed ropes around their necks. The four were standing before safe-deposit cabinets, the ropes knotted to handles of cabinet drawers; if gas fumes entered, Olsson was saying, the hostages would be strangled as they fell unconscious.

Olsson kept the hostages standing for hours with nooses around their necks. Kristin recounted that at this point, she "was numb with fright." She expresses the way in which her survival instinct is transforming into a kind of suicidal ideation, simply an intense desire for the fear and stress to end: "I wished I could just close my eyes, then wake up dead or alive." It's as though she is becoming a split personality, suffering from "borderline personality disorder." She has become two opposed selves with opposite agendas: one can live and one can die; it no longer matters as long as she can escape the unbearable stress.

The police made the decision to act and infused the vault with gas to save the hostages. As gas coursed into the vault, Olsson worked quickly to clear the door as he had an intense fear that gas caused dementia. The hostages were finally free. You might assume that their first act would be to run away, but here is how they behaved upon being liberated: "As they stood framed in the doorway, the convicts and hostages quickly, abruptly embraced each other, the women kissing their captors, Sven shaking hands with them. Their farewells over, all six walked out of the vault, Olsson and Clark in the lead."

Since the 1970s, we have had a psychological term that describes the way that powerless captives bond with powerful captors. In extreme situations, this mysterious and odd reversal can occur over six days. And yet we continue to be perplexed when children—who for *years* at a time are in the powerless "captive" position with abusive adults who are in the all-powerful "captor" position—feel a loyalty-bind with those who have abused them. Psychiatrist Bessel van der Kolk notes that children are programmed to *obey* and be loyal to the adults in their world. Thus, children can also suffer from cognitive dissonance and psychological trauma when they are "taken hostage" by adult abusers. This phenomenon may occur in sports where athletes normalize abuse and identify with the aggressor, an abusive coach, or in schools where students normalize abuse and identify with the aggressor, an abusive teacher.[2]

The psychiatrists who studied the Kreditbank crisis discovered that the hostages' brains were confused. Even after they were physically safe, "it turned out that they persisted in thinking of the police as 'the enemy,' preferring to believe that it was the criminals to whom they owed their lives." This is well-documented in Alex Renton's book on the elite schooling system in England, where abused children, once attaining adulthood, enrolled their *own* kids in the same schools they attended because they believed it was the "making of them."[3] Bonded to her captors and having severed ties with former attachment, "Elisabeth accused the doctors of seeking to 'brainwash' away her regard for Olsson and Clark." Yet, as time passed, the hostages, Elisabeth in particular, began to have nightmares and tap into their intense fear about the ordeal and her continued anxiety that the convicts might trap her again.

After the Kreditbank crisis, psychiatrist Frank Ochberg was galvanized by the psychological phenomenon of "Stockholm Syndrome." First, he developed a working definition for the FBI and Scotland Yard, and then, while working with the US National Task Force on Terrorism and Disorder, he put the syndrome into a series of psychological steps for them in case they faced any hostage-taking crises. The steps break down the reversal that occurs in the brain so that it is utterly confused.

> First people would experience something terrifying that just comes at them out of the blue. They are certain they are going to die. "Then they experience a type of infantilization—where, like a child, they are unable to eat, speak or go to the toilet without permission." Small acts of kindness—such as being given food—prompts a "primitive gratitude for the gift of life," he explains. "The hostages experience a powerful, primitive positive feeling towards their captor. They are in denial that this is the person who put them in that situation. In their mind, they think this is the person who is going to let them live."[4]

In February 1974, while Daniel Lang of *The New Yorker* was interviewing those involved in the hostage crisis at the Kreditbank in Stockholm, nineteen-year-old Patty Hearst was kidnapped from her apartment in Berkeley, California. Hearst was taken hostage by a group of armed radicals who called themselves the "Symbionese Liberation Army" (SLA). They broke into her apartment at night, beat up her fiancé, and threw her into the trunk of a car. Those who tried to intervene were threatened with gunfire. The SLA "wanted nothing less than to incite a guerrilla war against the U.S. government and destroy what they called the 'capitalist state.'"[5] They targeted Hearst because she was from a powerful and wealthy family and they wanted to use her for ransom. The SLA sought the country's attention and succeeded. They wanted food donations totaling millions of dollars. As the FBI reports, "They apparently began abusing and brainwashing their captive, hoping to turn this young heiress from the highest reaches of society into a poster child for their coming revolution."

Two months after her abduction, and after the beatings and rapes designed to brainwash her, Hearst announced she had a new name, "Tania." Her new name is indicative of the "borderline personality disorder" that emerges in the traumatized brain. Tania identifies with the aggressor and joins them in robbing a bank. Patty's life depends on Tania obeying the abusive, dangerous SLA. Forced into attachment fragmentation with all who love her and all she knows, captive Patty transforms into Tania, who aligns with the captors. It took sixty days for her to develop borderline personality disorder. Two months after her abduction, Patty/Tania was seen on a bank surveillance camera, wielding an assault weapon to protect and offer cover to her SLA captors. Her impulse to *protect* her abusers makes her a poster child for Stockholm Syndrome.

After nineteen months, Hearst and other SLA members were ultimately caught by the FBI and they were charged, including Hearst. While the grand jury conceded that she may have been taken hostage and then brainwashed by multiple beatings and repeated rapes, they still felt that she was responsible for her participation in robberies, wielding a weapon, making explosive devices, and being a fugitive from the law, along with her captors. They refused to see her as suffering from Stockholm Syndrome. From a psychological and neuroscientific perspective, Hearst was trapped in the cage of learned helplessness, under the traumatizing influence of her captors. The cage is constructed by our brain's will to survive. After physically escaping their captivity in the vault of the Kredit-bank, the hostages still believed the captors saved them and the police were their enemy. How long did it take, after a six-day trauma, to see clearly that the captors were the aggressors and the police were the rescuers? How long did it take Patty Hearst, subjected to her captors for nineteen months, to let go of Tania, put down the gun that probably saved her life, and return to a unified self? Were the FBI her saviors or the ones who put her in jail?

Survivors of child sexual abuse develop emotional "bi-directional" relationships with their abusers comparable to survivors of entrapment, with both suffering from Stockholm Syndrome.[6] Victims of sex harassment, as we saw with Ellen, do not want to report their abuser. If they do, they feel that they have betrayed them and harmed them. This guilty reluctance regarding their oppressor serves to "protect the abuser long after the abuse has ceased."[7] It answers the questions: If you were abused, why didn't you report? And if you were abused as a child, why did you wait until midlife to report? The answer lies in damage to the brain. The brain's defensive mechanisms, like those of a hostage, have gone into high gear and may not return to a healthy state without intervention. The traumatized brain struggles to distinguish between support and seduction, care and cruelty, motivation and manipulation. Bessel van der Kolk notes that the moment "we feel trapped, enraged or rejected, we are vulnerable to activating old maps and to following their directions." He works with traumatized patients to "observe and tolerate the heartbreaking and gut-wrenching sensations that register misery and humiliation." And he specifies that the harm has been done to the "warp and woof of brain circuitry" and has left scarring maps that are now "encoded in the emotional brain."[8]

Cross-species and evolutionary theory research in the animal domain reveals comparable brain-trauma reaction to entrapment and abuse as in the human domain. Researchers find that there is a neurobiological response in Stockholm Syndrome whether in monkeys or humans, whereby the brain's defensive mechanisms involve "paradoxically positive" relationships with their oppressors that may persist beyond release.[9] Studying behavioral patterns in animals reveals similar conduct as that displayed by someone suffering from Stockholm Syndrome. It hinges on "dominance hierarchies"; namely, who has the power? For the hostage, it is the individual with the gun; for the abused child, it is the adult in a position of trust and power. Victims go into defensive survival mode in both the animal and human realms: the brain strives to de-escalate the situation and offer reconciliation to the oppressor.[10] As we have discussed throughout, these brain distortions do scarring damage inside the skull.

These were deeply personal concerns for me because my sense of self was abducted in high school by three teachers who abused as a group. The grooming began at thirteen, but the intense abuse went on for a little over two years from sixteen to eighteen. I could have walked away during that time, but my brain was so confused, so loyal to my abusers that I suffered from learned helplessness. Who *taught* me I was helpless? My teachers. I wondered how the years of repeat traumas have impacted my brain architecture. There were many victims over the years. The teachers did not use guns to threaten us. They used humiliation or what psychiatrist Helen Reiss calls "weaponized unkindness," which is used for instance in cyberbullying to target the brain, not the body.[11] The teenage brain is highly susceptible to any kind of shaming because it is undergoing significant change that allows it to leave the parental "cave" and go out into the world and find a new "tribe." I use these ancient words because they explain the brain's shaping by evolution.[12] Regardless of our modern world in developed countries, the adolescent brain still develops according to the impulses and integrations of long ago. The three teachers took us out into the wilderness, which was an ideal cover-up for abuse, and it activated on a deep level the survival parts of our brains.

A child put into boarding school or taken out into the wilderness, where parents are far away or out of reach, may well suffer "attachment fragmentation" because their survival depends on identifying with the aggressor. As Bachand and Djak explore, this compares to Stockholm Syndrome: the child feels affection, loyalty, commitment—not to his family or friends—but to his captors, his abusers at school, in academics, in sports, in religion, or in the arts.[13] The child keeps these powerful adults' secrets. The child feels protective, perhaps even special at times. He is certainly dependent.

The power teachers have is that they assign value to you in the present and future. They assess how smart you are, what your potential is, what your character shows, what doors can open for you and what ones will close. Coaches have this same power to assess you as an athlete, give you opportunities to play or deny them, publicly celebrate awards you win or cover them up, write you an accurate letter of reference or damn you with faint praise, support your application for scholarship or block it at every turn, bring in recruiters or ensure they never see you play. Abusive individuals have immense power over what children love. Their power greatly increases with their ability to confuse what a child is passionate about with abusive experiences. What is loved becomes confused with what is loathed. My love of the wilderness was extinguished by these teachers, just as my son's love of basketball was ruined by his teachers who were coaching. After Ellen's suicide, I became concerned about the inability to escape the reversals. She had been physically safe, ensconced at university, a wonderful future ahead of her, but she remained trapped and confused in the past. Her "captor" ruled her brain. It occurred to me that I too—a middle-aged woman—had still not severed the loyalty-bind to the three abusive teachers. What would it take to do this act of profound disobedience? Why did my brain still operate in a state of fear that they could somehow harm me? The failure to save my son from abuse, the failure to save Ellen from killing herself, all hinged in some sickening way on my failure to save myself when I was a kid.

In 1974, when Daniel Lang was interviewing those involved in the Kreditbank hostage crisis and when Patty Hearst was being systematically traumatized by the SLA, two teachers began an outdoor education program in a high school in Vancouver, British Columbia, that they called "Quest." One teacher, Chris Harris, was nonabusive. The other, Tom Ellison, was a pedophile who for years used emotional and physical abuse to manipulate his many victims. In 2004, Ellison's victims *finally* got the police to take them seriously, and the police charged the former teacher, now in his sixties, with sex crimes. Harris, as is common in these abuse scenarios, came to the defense of Ellison. National news reported: "Chris Harris, who created the Quest program and taught it with Ellison between 1974 and 1977, says his friend is a man of 'absolute integrity' whom he can't imagine ever crossing the line. 'A man of morals,' he said. 'He's a very strong Christian, if that means anything. I just can't believe it.' Harris says he never witnessed or heard of any inappropriate sexual activity between students and teachers."[14] Little does Harris know that he is describing the Dr. Jekyll and Mr. Hyde split personality that fuels those who successfully abuse for years. "Dr. Jekyll" is a man of absolute integrity who doesn't cross lines, who is a man of morals, and who is a very strong Christian. Unfortunately, that doesn't rule out that he is also "Mr. Hyde," a man who sexually, physically, and emotionally abused students day in and day out, year after painful year. Just because someone is a talented Hollywood producer or a highly respected doctor for Olympic athletes or a Catholic priest does not mean that he would never harm someone and especially not a vulnerable child. In fact, extensive research documents it's far more likely that a successfully abusive individual fits the "pillar of society" profile described by Chris Harris. What's surprising is that we seem unable to identify how the split personality works in abuse cases.

After Harris left, Ellison was joined by another abusive teacher, Dean Hull, and then in the program's final years, a third abusive teacher joined them, Stan Callegari. Ellison was charged and convicted. Hull and Callegari have fifty victims' reports lodged with the police but have not been charged. While in 2008, the Quest story was called "the scandal that rocked the province," it was also described as an "open secret."[15] An aptly titled documentary film, *School of Secrets*, aired in the fall of 2007.[16] As per usual, schools and educational authorities "circled the wagons."[17] As one of the victims said: "It is disheartening to see that the VSB [Vancouver School Board, predecessor to the Commissioner for Teacher Regulation], continues to take no responsibility for its role in the abuse of so many."[18] At the trial for Ellison, there were those who said his environmental work "overshadowed any wrongdoing."[19] That's the equivalent of saying Larry Nassar's abuse was overshadowed by what a fine doctor he was or that Harvey Weinstein's abuse is okay because he was an exceptional film producer or that Larry Sandusky's sexual abuse of boys is excusable because he was a great defensive football coach. It's so indicative of the broken system and how it works.

At some point after Ellison was charged, I was called by national radio and asked if I would like to comment on my experience in Quest. Someone must have given my number to a journalist. I remember sitting in my office and speaking about the teachers. I said vague things about healing and moving forward and that people make terrible mistakes. If what I said could be summed up with one word, it would be "hazy." I spoke like I was in a haze because I *was* in a haze. But I'm not now. I wish that journalist would interview me now because I would tell them in no uncertain terms that what the three teachers did in the Quest program was massively destructive to the developing adolescent brains of not just the students who were sexually targeted but all the students forced into their psychotic domain of humiliation, threats, playing favorites, shaming, harassing, and overall manipulating. Hull admitted to the court that he had sex with an eleventh-grade student but went on to add that they had a relationship for ten years, which I guess meant to suggest that he "loved" her and wasn't a pedophile. He was middle-aged, married, balding; had three little children; and had not yet divorced his wife when he had sex with a sixteen-year-old student. I watched firsthand the grooming of this girl, and it was terrifying to watch.

At the outset, I was pressured repeatedly to engage in sexual relations with the teachers. I did not like their hugs or massages or comments or jokes. I wanted to be with my peers. On a canoe trip in the Yukon, I was constantly shamed for being "stupid," force fed, put down, blamed, ostracized, and ultimately abandoned in Dawson City for several days with next to no money for food. I was seventeen and did not even know where the teachers had gone. I had no way to contact my parents. My crime was that I had rejected their sexual harassment over the previous year. I did *not* want to slow dance with Hull. I did *not* want to give him a "blow job" for pulling me out of a rock-climbing impasse. I didn't even know at sixteen what a blow job was. I did *not* want to swim naked in front of them in the Yukon even though my peers did. It's almost impossible to say no to a teacher, especially when they publicly label you in front of your peers as "frigid" and "stupid." But finally, as I told the police, they "broke me" and I stripped off my clothes and had to endure them commenting on my naked body. That was the only moment in my two-hour-long victim impact statement to the police that I cried. I also asked the police to apologize to the two girls who were drawn into sexual relationships with Hull and Callegari that seemed solidified on that ghastly trip. I felt guilty for having been used to solidify their dependence on the teachers. The more kids like me were humiliated, the more others were taught to obey, succumb, give in, survive. There are few things that threaten the adolescent brain's survival more than public humiliation. It is a powerful weapon.

The part of my brain tasked with survival assessed the trip in the Yukon as a full-on threat to my life. While my prefrontal cortex, not developed nor mature at seventeen, did not understand the complex manipulations of middle-aged men, deeper brain structures were on high alert. While my brain had no words to report to my parents—or articulate even to myself—that I was being abused along with my peers, the trauma-recording apparatus in my brain was working double time.

As students in the wilderness, we had no way to survive without Hull and Callegari. They knew the weather, the river with its white water, the routes through hundreds of miles, the correct gear to bring. They carried shotguns in case of grizzlies. Far from our parents and civilization, they could do anything to us they wanted. Six students were brought along as cover to make sex abuse with two teens seem like a "school trip" in an outdoor education program. We never breathed a word about it to one another. There were no words for it. We were simply repeatedly shocked with the aberrant behavior of our "teachers" who, since the age of four, we had been trained to obey. This is why I don't fall for the "true love" story they peddled to authorities and the community when they were at risk of criminal charges like Ellison. They didn't need to worry about their victims reporting because these two teen girls wanted to protect their abusers. They were bonded to them. They felt great affection, perhaps even love for them. Is it possible they were suffering from Stockholm Syndrome? When multiple victims finally had their abuse taken seriously by police, Ellison was living with a former student and had had a child with her. Was it "true love" this time? This former student, mother of his child, defended him throughout the trial.

I was developing a borderline personality disorder and didn't even know it. For years as I saw excellent psychologists and psychiatrists, I never once reported any of the abuse I suffered in the Quest program. That was not "me." It happened to someone else, just like Patty was abducted, but "Tania" was protecting her captors with a gun. The trusting, confident, healthy girl I once was had been severed, and in her place was a hardworking, distrustful, deeply insecure, unhealthy woman, a woman who compensated for intense feelings of self-loathing by overachieving in the academic and professional world. While I was in a haze for almost forty years, which is textbook since I was suffering from Stockholm Syndrome or at least a loyalty-bind, the realization that what had happened in Quest was deadly for the brain hit me in the gut when reading an article by a former Quest student in *Elle* magazine. The title was "Learning Curve: Sex with a Teacher." The author, Katrina Onstad, promised in the subtitle to inform her readers about "*What's really going on when girls hook up with their teachers.*"[20]

My panic attacks always began with a trigger: reading the Commissioner for Teacher Regulation's reports saying the teachers who abused our son and so many students would not be held publicly accountable; seeing a teenage girl waterboarded in a TV show; reading a novel where children were pitted against one another to serve adult needs. I would feel energy coursing through me, making my hands shake. Then I couldn't breathe. I'd feel a combination of rage and powerlessness that I could only express by scratching my forearms until they bled. Just the title of Onstad's 2007 article in *Elle* was enough to trigger an attack.

Now middle-aged, Onstad talks about her time in the Quest program, several years after me, and her disappointment at not being specifically targeted for sex abuse. Onstad faults her appearance for not being one of the lucky ones targeted by pedophiles: "Nothing had happened to me. With my androgynous looks and gangly body, I was hardly babe enough to have attracted his attention, and that year, I was untouched, and in a way, largely unnoticed." The article details Onstad's vivid memories of scenes of public humiliation and the way the teachers favored certain girls, who then became their sexual partners. Sad not to have been on such a "learning curve" with her teachers, Onstad describes Ellison, Hull, and Callegari as if they were alluring figures.

> We were rich kids who had skipped through life in a haze of clueless entitlement, and we deserved our comeuppance. One day, one of the three wheeled a cage that had been holding volleyballs and basketballs to the front of the room. A boy was put inside the cage in either his gym shorts or his underwear—it's unclear to me now. But I do remember Stan and Dean laughing as they told him to suck his thumb like the little rich baby he was. It was Tom who picked apart our bodies: "Go fat-ass!" I remember him yelling at a girl as she ran sprints. During the trial, an ex-Quester testified that the Boys would pinch girls' breasts and slap their butts.

Does this sound like Stockholm Syndrome to you: *we deserved it?* This is the mantra abusers hammer into the heads of their victims and very few have the mental strength to push back until decades have passed. Onstad does not clarify how relentless emotional, physical, and sexual abuse was a fitting way to punish children whose parents had money, but she is clear the abuse was deserved.

The article is extremely well-written. It is honest in its portrayal of the Quest program to my mind. What disturbs me, however, is that the author ends with victim blaming that she links to her own daughter: "So now we watch our daughters, pretending that we can save them from themselves. As if they won't find their own ways into the forest." "Forest" is the metaphor for sexual relations with teachers, and Onstad suggests our daughters cannot be saved because they're asking for it. We can't save them "from themselves" because they will find their own way. It's an odd blend of giving children freedom to have experiences, explore, discover who they are, and believing that somehow this has to occur through sexual abuse. That's the amazing lie that abusers tell to children: "You cannot survive, perform, have opportunities, meet the right people, play your sport, go into the wilderness, achieve academically without me. So do as I say and I will ensure you get to do what you love and succeed." And some believe this into adulthood. It's quite remarkable to read an article by a mother longing for her daughter to be sexually abused, but it is also indicative of how abuse works on brains. Onstad might have suffered more than she realizes. I believe her brain is showing signs of Stockholm Syndrome. Onstad claims she was more aware than I was as a teenager: "Quest was definitely a mental torture chamber, but the word 'cult' isn't right; it presumes we weren't aware." I would beg to disagree and feel that Onstad herself has never left the torture chamber or the cult-thinking of the Quest program.

It's worth looking more closely at how Onstad explains her sadness that her own daughter won't ever be accessed by the likes of Tom Ellison.

> My daughter will never do a program like Quest because it could never exist today. We will design our lives to keep her safe from all risk: emotional, physical, and sexual. She will be electronically shadowed and chauffeured, all precautions taken to keep her away from threats like Tom Ellison and the great unknown. Such private experiments on the way from girlhood to womanhood won't be available to her. I suppose I should be comforted, yet I can't help but think that with all this safety comes a great loss.

Onstad refers to articles by professor of sexuality Deborah Tolman about teen girls and their sexual desires and throws in professor bell hooks, who writes about the "erotic energy" between teachers and students. She turns to anthropology and initiation rites to position the experience as somehow liberating and feminist.

But she does not seem to have done any research on sexual abuse in the past or in the present. This must be why she appears unaware that rampant abuse in all forms did not stop in 1987 when the Quest program was finally shut down.

Onstad's daughter can *still* seek out sexually abusive adults in schools, clubs, sports, places of worship, the workplace, and the arts. The list of opportunities is endless in the broken system that responded to ten years of abuse in the Quest program with the idea of creating a new "code of conduct." This is exactly what the headmaster decided must be done at my son's school when rampant abuse was uncovered thirty years later: he set about creating a new "code of conduct," which was odd because he also maintained there hadn't been any abuse. In 2019, the Honorable Minister for Science and Sport, Kirsty Duncan, announced that a new "code of conduct" would be developed when she was confronted with rampant athlete abuse in 2019.

It's time we exited the bullying paradigm. It's time we cut through the Stockholm Syndrome that leads parents like Onstad to believe that the abuse in the Quest program must be preserved so her daughter can experience it. It startled me to realize I too was suffering from Stockholm Syndrome. I hadn't just kept the secret of the three abusive teachers for forty years. I had been protecting them.

# Step 7: Believe in Yourself

I am a reader and a writer of culture. The way I snapped myself out of Stockholm Syndrome was by reading everything I could find about the Quest abuse, Ellison's trial, Hull and Callegari's avoidance of charges, what principals had done, what the school board covered up, and what other victims or nonvictims reported. I reported to the police, which did not result in charges despite them telling me they had fifty victim statements. Maybe it's because the teachers are now in their late seventies and eighties. I spoke to other victims with whom I had shared traumatic experiences and heard their perspectives. I spoke with friends and colleagues. At first people said: "Oh, what happened to you wasn't that bad. It was far worse for me or for someone I know or back in the day." And then when I narrated what happened in detail, they said, "That's not believable; no teacher could get away with that."

And then I started writing. Mostly the writing has been therapeutic, but some of it I have published as guest blogs for *Kids in the House* and *Edvocate* because I feel strongly that we set children up to be victims by failing to educate them about how abusive adults operate. Every kid in kindergarten, or earlier, should be instructed in detail about grooming, luring, sexual harassment, sexual abuse, who to report to, how to narrate what's happening with the correct terms, what to do if the adult to whom they report responds with disbelief or concerns about the perpetrator's reputation or suggests to the child he or she might be lying. Every child needs to know what verbal, psychological, and emotional abuse and emotional neglect are and what to do if they are a victim, likewise for all forms of assault, physical abuse, and physical neglect. Children need to know that emotional and physical cruelty often sets the stage for sexual abuse, as we learned in detail from the victims of abuse in USA Gymnastics. Children need to understand gaslighting and how manipulative it is. This knowledge is far more important than math or science or music or language. They have a right to know what all forms of abuse do to their brains and especially the rapidly developing brain from zero to three and thirteen to twenty-four. This can't just be a one-day workshop or a pamphlet. The brain learns by repetition at timed intervals. Children need to be taught about abuse as often as they're taught all the other skills we believe are important like soccer, math, swimming, drawing, reading, chemistry, history, and so on. Knowing how to recognize abusive conduct and halt it is arguably the most important skill a child can develop to lead a healthy, fulfilled, and happy life. Children must know who to turn to for help and ensure that the adult or agency does not have conflict of interest. As argued by law professor Amos Guiora in his recent book *Armies of Enablers*, those who receive reports of abuse and fail to protect victims must be criminally charged. If we want to see an end to institutional complicity, then there must be consequences for aiding and abetting child abuse. Child safety needs to be something that is seen as a sacred trust.

When I began writing about the abuse I endured as a child, people told me *not* to use the teachers' names. That's when I truly recognized how Stockholm Syndrome does not only scramble the brains of victims; it infuses the whole bullying paradigm. The moment you record the names of those who have abused you is the moment you take your power back. Research shows that "naming an affect soothes limbic firing." As psychiatrist Daniel Siegel explains, sometimes "we need to 'name it to tame it.'"[21] He advises to use the "left language centers" of the brain to "calm the excessively firing right emotional areas." I took this advice, but the naming for me was literal. I wasn't naming the "emotional dysregulation," the "overactive feelings," the "affect," as though they were problems with me or my brain; I was recognizing that they were *outside* of me. The feelings my abuse generated in me were laid consciously at the feet of the teachers who harmed me and so many others. No more blaming and shaming myself. The accountability lay with them, and they were separate from me. I refused to identify with the aggressor. Putting the abuse they did into language served to do what Siegel said it would: the imbalance between the "emotional-generating subcortical areas" and the capacity of the "mental distance," the "sanctuary" created by the language-based area of the brain, once again came into alignment. This healed my brain, which previously found itself "overwhelmed by fragmented autobiographical images, filled with bodily sensations, awash in emotions that overwhelm and confuse." Doing the hard work of facing the abuse done to me, naming the abusers to tame the harm done to my brain, returned to me an aligned, holistic Mind-Brain-Body.

The two vital principles that the bullying paradigm cannot bear or support are transparency and accountability. Identifying *who* the aggressor is means you don't identify with the aggressor. Identifying who the aggressor is ousts the Mind-Bully right out of your brain and opens the door to creating an Empathic-Coach. Psychologist Angela Duckworth says that learning the key quality of hope means sometimes you need to "*ask for a helping hand*."[22] When you've been abused by those who were empowered and entrusted to help you, it's possible to fear seeking help. That's why it's key to know very clearly that those who abused you are fully separate from you and essentially random figures in your life. If they betrayed their trust, if they failed in the sacred duty of caring for a child, well, that has nothing to do with you. Walk away. Remain crystal clear that their abuse targeted you, like others, and none of you bear any responsibility for adults' abuse. You have every right to ask for a helping hand. You can reach out and take the hand of the Empathic-Coach.

Daring to say the name of the aggressor means that you greatly increase the chances of the aggressor being held accountable on one level or another, even if it's only their own conscience. The failure of the bullying paradigm to hold abusive individuals accountable is tragic because it stops them from seeking help; it blocks them from brain rehabilitation; it results frequently in more victims, which is by far the worst and cruelest outcome. Exiting the bullying paradigm means refusing to keep quiet about who abused you and what they did. It means that you refuse to feel any shame or self-loathing. Instead, you lay responsibility squarely at the feet of those who did the abuse. You refuse to keep the secret of their abuse.

Brace yourself for the abusers' private promise to change because they did not know their conduct was harmful, followed by a public denial, aggression, threats, and personal attacks because those who are abusive and those who support them are well-trained in aggressively pushing back. They've had to do it before, and they're quite ready to do it again. Any kind of criticism or questioning of their conduct results in a grand display of outrage, hurt feelings, and shock, which then turns into accusations, a well-honed victimhood, and then claims they are the innocent target of a witch hunt. It's textbook, but it's not a bad idea to prepare for it. If you don't want to waste your time and energy with this well-scripted routine that occurs regularly in the bullying paradigm, then just write—just name it to tame it—for yourself to make sure that you do not suffer from Stockholm Syndrome. I prefer writing, but you may use any method you like to unpack the psychological and brain-scarring impact of trauma from abuse. The key is to become very clear so that you are not identifying with the aggressor.

Before I could commit to my own healing and restoration of health, I had to oust the abusive teachers from my brain. I needed to free up cortical real estate if I was going to do brain training. I needed to consciously heal my borderline personality disorder. My intellectual brain—out performing in the world—was disconnected from my traumatized, emotional brain. My intellectual brain was dedicated to high achievement in academia, namely pleasing "teachers" and acting as though their extrinsic valuing of my accomplishments would make me healthy and whole. It had the opposite effect. It divided me further and further so that my holistic Mind-Brain-Body was fragmented and acting at cross-purposes. The emotional part of my brain was full of anxiety and fear and on the alert for danger. It was not a matter of needing a ladder like the kids in Stronger Brains; I needed a bridge. The two parts of my brain—intellectual and emotional—had a chasm between them.

Into all of this distortion and noise in my brain came the calm, knowledge-able words of Merzenich: "It is tremendously destructive for you to tell yourself that 'you can't,' if, in fact, succeeding just requires a little more serious effort and practice on your part. Your brain registers all of that negative messaging. When you tell your brain 'I can't' just a little too often, it becomes a self-fulfilling prophecy."[23]

Step 1 is to identify the reversals of Stockholm Syndrome by naming and taming the abusers. Write on paper, say out loud, put into a painting or a play these words: I am not the protector of those who harmed me. I am their victim. I do not need to cover up what they did because I was victimized. There's no shame or self-loathing or fragmented sense of selfhood. Underlying those pro-jected falsehoods is my whole and healthy brain waiting patiently to be retrained back to its glorious potential. It takes work, just like getting the body in shape takes work, but it's absolutely doable. I believe in myself and my capacity to harness my neuroplasticity to heal my scars and restore my health. If my Mind-Brain-Body has been broken asunder, now is the time to fix it.

Step 2 requires you to believe in yourself. It requires you to tell your brain "I can" and really mean it. Step 2 paves the way for you to commit to the deep practice of brain training, empathy, mindfulness, and aerobic exercise, as we will look at in the following chapters.

These two steps brought my panic attacks to an end and set me on a path to holistic wellness of a kind I had not experienced since before the abuse began. I opened up Merzenich's brain-training program again. For the sake of research, it asked why I was doing the training, and I selected "recover from injury or condi-tion." I was recovering from the injury of bullying and abuse. I was recovering from the condition of aligning with the very individuals who hurt me as a child and traumatized my brain. It was that condition that I needed to be rid of once and for all. It was amazing to me how much I wanted to earn the "stars" in the program that allow you to advance to the next level. It was a little daunting but inspiring to be able to chart my own progress. I knew that changing my brain and improving my cognitive health was going to be as hard work as rigorous daily exercise, but in our quick-fix culture, it still surprised me how much I wanted instant results. While there were exercises for "attention" and "processing speed," I started out with the exercises to strengthen and heal "memory."

# CHAPTER 8

# Heal Your Brain with Your Mind

## *Step 8: Redraw Your Brain Maps*

My son Angus walked carefully into the kitchen, and I could tell that his pain levels, normally high, were off the charts. When his pain is uncontrollable, his shoulders curve in; he becomes very pale, keeps his eyes down, and has lines etched into his forehead that more often are seen on someone who's middle-aged as opposed to twenty. I asked the obvious question: "Are you in a lot of pain?"

He nodded and went over to the sink where he filled up a glass of water. His movements were slow and deliberate as if any sudden movement might make his head explode. He started toward the stairs.

"Did you take codeine?" I asked. Sometimes he can head off the worst of the pain by taking four codeine pills.

"Yeah, but it doesn't do anything." His voice was strained as if speaking made it all worse.

"The Botox isn't kicking in?" A week previously, the anesthesiologist at the hospital had injected the area around the top part of his neck with Botox, as months earlier, the first injections had lowered the pain levels in his neck and spine.

"Not yet," he replied.

"It's taking so long this time," I said, again stating the obvious. I felt power-less as per usual when faced with his debilitating pain.

He paused at the foot of the stairs as if walking up was a strain he couldn't handle. He turned and met my eyes.

"I wish I could take your pain away," I told him. "Take it away and absorb it into myself so that you didn't have to feel it."

Unfailingly honest, he replied, "You couldn't handle this kind of pain. You wouldn't be able to read or write or even think. You wouldn't be able to stand the sunlight."

He went slowly up the stairs, one foot after the other.

Four years earlier, in November, Angus dropped out of school. He had undergone a second corrective surgery on his legs and was overwhelmed by the pain and apparent misery of his situation. He had been managing a whole series of health issues that he had been born with and that seemed disconnected. He had celiac disease, kidney disease, cleft palate, and ear-nose-throat (ENT) problems. He had undergone multiple surgeries starting when he was a baby to deal with these medical conditions. That November, he dropped out of school and retreated to his room. He kept the blinds down and noise-canceling headphones on. We were lost as parents. It wasn't until January that he was diagnosed with Klippel-Feil Syndrome (KFS), which is a rare genetic disease. In utero, the spine does not separate out all the vertebrae properly and Angus's top two, C1 and C2, were fused. This wreaks havoc on the body, resulting for many in kidney disease, ear-nose-throat problems, cleft palate, gastrointestinal issues, and severe, persistent, neuropathic pain. We had a handle on these medical challenges, but what we hadn't been prepared for was the onslaught of mind-blowing pain that began with his growth spurt in adolescence.

Angus dropped out of school because of the physical pain but perhaps more because of the social pain. While a number of teachers who worked with him were incredibly supportive, there were some who did not understand just how physically compromised he was. On one particularly awful day, a teacher called him "pathetic" in front of the other students as he sat hunched over his laptop unable to participate in gym class. On another day, this same teacher pursued him to my office. I was in another class at the school teaching. She put her hands on either side of my office chair where he was sitting, got up in his face, and said, "You're not trying." Ultimately, kids began to echo these adult comments, and Angus was bullied as being a "f—ing fake and a faggot." That teacher brought tears to Angus's eyes, as did the kids who turned away from him. No amount of stressful or painful medical intervention ever made him cry, but not being seen and understood, in terms of his relentless pain, brought tears. My son's intense suffering is described in psychologist Matthew Lieberman's work on how wired our brains are for social connection. His research using fMRI shows that we "intuitively believe social and physical pain are radically different kinds of experiences, yet the way our brains treat them suggests that they are more similar than we imagine."[1] While Angus could handle the physical pain, the social misunderstanding, that morphed into bullying, was simply too much.

His savior, along with remarkably dedicated medical personnel, from surgeons to massage therapists to a gifted osteopath, has been mindfulness. I don't think I would have believed in the incredible promise of mindfulness practice, if I had not witnessed it firsthand in a child with chronic pain. Doctor and mindfulness teacher Amy Saltzman has done a large clinical trial on the benefits of mindfulness for those suffering from chronic pain and illness. Her definition of what mindfulness is applies to my son's use of the technique. "Mindfulness is paying attention here and now, with kindness and curiosity, so that we can choose our behavior."[2] For Angus, mindfulness lets him *choose* life over death, constant pain over escape. He daily makes this heroic choice. Targeted mindfulness can heal the brain and reduce, sometimes even remove, pain felt in the body, and it can also remove pain felt in the spirit. Angus learned mindfulness techniques as a child attending "Kids in the Spotlight" at The Haven, which is a center for personal and professional transformation on Gabriola Island, off the coast of Vancouver, British Columbia. The program was designed and run each summer by developmental psychologist Denise Goldbeck. Starting when he was eight years old, Angus learned that breathing and visualizing in combination could lower stress levels, soothe anxiety, and calm his nerves. Little did he know that as he grew up, these techniques would help him survive countless surgeries and ultimately unbearable, relentless pain.

When the pain overcomes even his well-honed ability to fight it, he lies down and begins box breathing. He takes in four slow breaths, holds it, then releases for four and again holds it. While he does this slow, purposeful breathing, he imagines a place where he feels safe, seen, and joyful. That place is The Haven, peopled with young people who have become cherished friends.

In his late teens, Angus moved from "Kids in the Spotlight" to another Haven program called "Teens Alive," led by transformational counselors David Raithby and Linda Nicholls. In this program, that supports the initiation of youth into adulthood in healthy ways, Angus encountered teens who did not have his physical suffering but more than matched his pain with emotional pain and suffering from abuse. He is a highly empathic individual whose heart nearly collapsed at the horrendous experiences some of the teens had with abuse in their childhoods. Again, mindfulness was crucial to the healing and survival of these young people, and it is well-documented in the scientific research to have these healing properties.[3] All of us have within us a remarkable capacity to heal trauma to our brains and the pain that accompanies it through embodied breathing and mindfulness practice.

For those who are unfamiliar, mindfulness practice—which is all about *how* you pay attention—is a cornerstone of Eastern health practices and has been for close to three thousand years. In the twenty-first century, there is extensive neuroscientific evidence that shows increases in cortical thickness and hippocampal volume, along with a reduction in worry, anxiety, and depression, in subjects who commit to merely eight weeks of mindfulness-based stress reduction programs.[4] MRI brain images were taken from sixteen healthy, meditation-naïve participants before and after an eight-week program. Results showed that slow breathing, focused on deep, calming breaths for twenty-seven minutes daily, increased the density of gray matter in the hippocampus, which is a part of the brain involved in learning, memory, self-awareness, compassion, and introspection. Not surprisingly, with all that gray matter getting denser, participants reported an increased sense of well-being and peace.[5] Compare this healthy, thriving hippocampus to John Ratey's description of a hippocampus constantly bathed in the stress hormone cortisol, appearing on brain scans as "shriveled" like a raisin.[6]

Mindfulness enhances "reflection, relationships, and resilience." Research is clear that mindfulness practice enhances social and emotional intelligence. Not only is mindfulness exceptionally good for your brain, but these practices have also been shown in multiple studies to support a healthier body.[7] Studies show that mindfulness reduces stress and helps alleviate many psychological and physical disorders, especially those involving anxiety, trauma, and addiction.[8] Studies reveal that when you are happy and relaxed, you are able to focus on the big picture and do well on tasks that require memory retention.[9] In contrast, those who are anxious and stressed, afraid of being humiliated, hone in on small details oftentimes associated with a dark emotional state, such as feeling shame. When you're stressed and anxious, you forget most of your learning and struggle to transfer knowledge to new situations. As shown on MRI, SPECT, and EEG scans, mindfulness triggers "positive neuroplastic changes" in your brain that manifest as better health, sleep, memory, concentration, and mood, not to mention an enhanced ability to control conduct and emotions.[10]

Working with executives and others suffering from high stress, neuroscientist Stan Rodski discovered that encouraging his patients to color in an abstract drawing, combined with mindfulness over a number of sessions, was highly effective in helping them rewire their brains. He worked with a teenager suffering from extreme anxiety, and he asked her to do an experiment with him. He connected her to an EEG so that she could see on the computer what was happening in her brain before and after she combined purposefully coloring in the abstract drawing with doing mindfulness practice. Rodski reports that his young patient was amazed to feel the dramatic improvement in how her mood felt but also to see the results of her lessened anxiety on the computer-screen image of her brain. Over the course of several sessions, she moved from using very dark colors, indicative of her stress, to brighter, more "light-hearted" colors that conveyed her more calm and mindful state. This shift was empowering for her as she realized that *she* could make conscious brain choices that would then affect her unconscious brain states that were producing the extreme anxiety.[11] "Unconscious brain states" compare to the Mind-Bully that we might not even know is there, talking at us, putting us down, limiting our unlimited potential. If we do not become conscious of the brain states or the bully in our head, we may struggle to seek cures and remedies such as mindfulness.

While many of us know that, like physical exercise, mindfulness is incredibly healthy and life-enhancing, nonetheless we feel blocked and fail to commit to a simple half an hour per day. How is it possible that we continually circle back to decisions and activities that can ruin our relationships, rob us of health, and endanger our very lives? It's pretty amazing when you think about it, but when you look into your brain and think about how it may well have struggled in our society—riddled with bullying and abuse—it's not all that surprising. What we first have to do is question the bullying and abuse indoctrination and gaslighting and then replace it with a more mindful, insightful, and healthy approach. It's indoctrination that has likely occurred over the course of years, so it cannot be easily or quickly dismantled. It takes significant effort and belief that your Mind-Brain-Body can achieve freedom from the cage of learned helplessness. Part of succeeding at exiting the bullying paradigm is understanding how it sets us up to fail. Thus, we have to take some more time examining just how much it has infiltrated our brain before we can truly set in motion an effective remedy.

We have discussed the dangers of toxic environments and chronic stress, which activate the sympathetic nervous system. This is how our brain responded to stress back in the day, when there were predators like massive elephants and tigers roaming around, and funnily enough, it hasn't changed much. While our brains are ridiculously complex and sophisticated, they are simultaneously stubborn and old-fashioned on the issue of how to cope with stress. The brain prefers a straightforward and simple response. It doesn't matter to your brain that your stress is a public-speaking engagement or an overdue credit card payment, it interprets *all* stress as a predator, and so it prepares you to fight the beast, run away from it really fast, or freeze in place hoping you blend in with the grasses on the Serengeti. It pumps up your heart; it ramps up your blood pressure; it blasts adrenaline and cortisol into your system as they are designed to help you survive. The problem is that you don't need to fight the public speaking engagement or run away from your debts. All of this well-meaning stress response, especially when it isn't a quick survival strategy that subsides in a half an hour or so, is counterproductive in the modern world. We live with a ton of stressors that aren't life-threatening and that don't require massive physical action, but our brains do not change quickly. Therefore, in the twenty-first century, the fight-flight-freeze response is taxing on your brain and body. As discussed earlier, that's the bad news, and if you read the neuroscientific research on it, it's pretty grim. But here's the good news. What calms down the sympathetic nervous system? You got it: mindfulness.

Mindfulness practice sets in motion our parasympathetic system, which is the opposite of our stress response. The parasympathetic system calms down the racing heart, lowers the blood pressure, and replaces the stress hormone cortisol with the neurotransmitters dopamine and serotonin. It relaxes your brain and body, which helps you to recharge, keep perspective, and heal (both emotionally and physically). It returns to you a sense of control, which is the opposite of learned helplessness. Mindfulness helps you to be relaxed and not lash out at others. It leads to responsive, as opposed to reactive, conduct. It allows you to develop resilience, and it enhances your creativity and concentration.[12] Pretty amazing if you think about it. Mindfulness is a game-changer, and it gives you a competitive advantage, just like happiness does. Positive psychologist Shawn Achor references the research on the way in which meditation has been shown to grow the left prefrontal cortex, which is the part of your brain that's engaged in "feeling happy."[13] Achor explains that the daily practice of slowing your breath and staying focused in a mindful way can "permanently rewire the brain to raise levels of happiness, lower stress, and even improve immune function."[14] Having a bullied brain can result in an unhappy disadvantage. Mindfulness is one of the ways to rewire neural networks to obtain Achor's "happiness advantage."

Learning, through sustained activity, lays down maps in your brain.[15] You can unlearn a map that has become painful or no longer serves you. With mindfulness, you can choose flow. Amy Saltzman describes the flow from mindfulness as "that moment when time slows down and the world disappears, when even the sense of 'you' as a separate self falls away and there is just movement, rhythm, energy, joy."[16] I took a workshop with Saltzman and later interviewed her. When she speaks, you can literally feel her kindness and curiosity as she carefully chooses words to share her wisdom. When I spoke with psychiatrist Daniel Siegel, another mindfulness expert, what struck me was how his kindness and curiosity showed up in the form of intellectual and social delight. He said that his mother had been reading his book *The Mindful Brain* and told him that she realized that "mindfulness was about turning annoyance into amusement."[17]

Mindfulness or meditation teacher George Mumford is usually photographed sitting cross-legged, and surrounding him in a circle, eyes closed, lying on their backs in basketball uniforms, are the elite athletes he has practicing mindfulness to reach peak performance, elite athletes like Michael Jordan and the late Kobe Bryant. The combination of mindfulness with sport and fitness is pretty much unbeatable in terms of creating healthy, high-performing brains and bodies. You can use your mind to heal your brain and body as elucidated in the work of cardiologist Herbert Benson, founder of the Mind/Body Institute in Boston. Research shows that the genes being changed by mindfulness and relaxation techniques are the very genes "acting in an opposite fashion when people are under stress."[18] Neurologically speaking, mindfulness translates as "tamping down your own amygdala-driven threat sensors" while listening to your inner voice or to someone else (like a yoga teacher or mindfulness expert) who is speaking.[19] As you recall, the amygdala is your brain's alarm center (among other things), and it can be highly reactive if you've been subjected to bullying and abuse. It might have a tendency to overreact and see everything as a threat and thus distort and disrupt your relationships at home and at work. This is where mindfulness can be incredibly helpful. It replaces the panicked messages from the stressed amygdala with calm, centering, kind, bringing-into-the-present messages. It is a conscious choice to decide there isn't something life-threatening in your vicinity that makes you want to fight, flee, or freeze as per the sympathetic nervous system. You can be purposeful about calming down and tackling any crisis in a mindful way that allows you to activate your parasympathetic system, which is way better at problem solving, articulating the issue, being empathic, being creative, and being self-compassionate.

Like Stan Rodski, George Mumford discusses the way in which the parasympathetic system, set in motion in the brain and body through mindfulness, is the opposite of our stress-system or sympathetic system (fight, flight or freeze): "Instead of speeding things up and flooding us with stress hormones, it actually slows us down. It lowers blood pressure and slows our heart rate." The parasympathetic nervous system, activated by mindfulness, releases "a neurochemical called acetylcholine," which is a vital ingredient in the relaxation process of the brain and body.[20] Neuroscience is catching up with what the ancients have known for thousands of years. Intentional deep "breathing can even have an impact on our genetic makeup on a cellular level."[21] If you are starting to feel the early symptoms of stress rising up in your system, creating anxiety, increasing your heart rate, and raising your blood pressure, you can take control and calm the whole episode down by slow, intentional breathing, channeling mindful meditation. You can drop your twenty-first-century anxious ego-state and replace it with deep, unconscious knowing of the calm that resides within and is perennially available to you. Discovering that you have this power over your stress and emotional states can be transformative.

As outlined by Stan Rodski, the five key aspects of mindfulness are:

1. paying attention to your breathing, to your body, or to an activity like walking, coloring, or knitting;
2. paying attention in a deliberate and focused way so that random thoughts, feelings, and distractions are gently dismissed to maintain focus;
3. paying attention with the express purpose of calming down the brain and body;
4. remaining in the present, gently dismissing memories or concerns about future duties or worries; and
5. being nonjudgmental or having self-compassion throughout the process.

The key takeaway to remember is that mindfulness is a process.[22] We have a tendency to think that after we've tried something a few times, we should be good at it. This can be the same quick-fix thinking that derails you from an exercise program or from Merzenich's brain training. They're all a process. They're all hard work, but each one is worth it.

I can tell you from experience that mindfulness sounds deceptively easy. The challenge is to make your mind understand that this is transformative for your brain and body. It's a priority, and it can liberate you from stress and learned helplessness. As you know, when you want to do healthy acts that promote your well-being and happiness, you trigger all too often the Mind-Bully who convinces you that it's too hard; it won't work; it's all just a conspiracy to make you hopeful when in fact you'll never be whole, healthy, or happy, so why keep trying? The Mind-Bully's mantra is scripted to use all possible avenues and encoded maps to erase your belief in yourself. You need to rise up against it quite consciously or it can infiltrate your mind. That voice, those limiting beliefs, those entrenched neural networks are extremely hard to resist and to reject. To douse their fiery rhetoric, we need to examine how they came to rule in our skulls in the first place. Once identified, it makes it a lot easier to ignore them and replace them with mindfulness and empathy. If they still pose a problem for you, you might want to reread the earlier chapters in *The Bullied Brain*. Neural networks that were laid down over years won't be easily rerouted. It takes time and practice. The brain learns not by reading something once but by repetition at timed intervals.

In an innovative move, NBA coach Phil Jackson hired George Mumford to be the mindfulness coach for his professional basketball players. Jackson wasn't your usual coach. He created for his team an "inner sanctum" for them to meet. He decorated the meeting room with Indigenous totems and other symbolic objects. He has a special relationship with the Lakota peoples, who named him "Swift Eagle," an apt description of his own playing style on the Knicks that landed him not one, but two NBA championships. Jackson shares: "I had the room decorated this way to reinforce in the players' minds that our journey together each year, from the start of the training camp to the last whistle in the playoffs, is a sacred quest."[23] That's how you should see your own journey. Instead of lamenting how difficult it is to do daily brain training or practice mindfulness or get fit, channel Jackson, and recognize that although it's hard, you are on a "sacred quest" to reunite your Mind-Brain-Body to optimize your healing, health, and happiness.

Professor Lori Desautels and educator Michael McKnight note that in the "Lakota language, the word for a child is 'wakanjeje.' It literally translates as 'sacred being.'"[24] Along with Phil Jackson or "Swift Eagle," we should try the approach of the Lakota people and see children and treat children as if they are sacred beings. Your inner one-year-old, eight-year-old, and teenage self are all sacred beings, and when you begin to see the inner sanctum of your skull with this same reverence (remember it's the site of eighty-six billion starry neurons), then you will find the fellowship you need to set out on your sacred quest for wholeness. One of the most effective ways to exit the bullying paradigm is to understand that children, sacred beings that they are, haven't been indoctrinated. They have not given up their critical thinking to simply obey. They have not distorted their thinking to believe that abuse made them successful or even great. Children should not be "taught" bullying by adults. Children *are* the teachers. With their brains wired for empathy, it is children who can lead us out of the bullying paradigm and into a new healthier, happier paradigm.

Mumford works with athletes on kinesthetic imagery or visualization so that the body experiences movement via the mind. The athlete *mentally* rehearses the goal she wants to achieve, and using "outcome expectation as the frame, we actually rewire our brains to reflect the activity as if we were really doing it."[25] This same technique can be applied to other learning such as when someone who bullies or abuses consciously works to rewire their brain by imagining treating their victim with empathy and compassion. Likewise, an individual who has identified with the aggressor and developed a Mind-Bully can consciously shift the bullying and abusive messaging or behaviors into those of an Empathic-Coach. Unwiring destructive thought patterns or entrenched mindsets, along with the behavior that accompanies them, is not easy. It takes effort, work, daily practice. As Mumford finds, even with elite athletes at the highest level, it takes concentration and work, but he adds that it's worth it because research shows that athletes "who can deliberately keep their mind calm and focused tend to perform at or near their personal best."[26] Remember this research when someone in the bullying paradigm tries to insist that athletes, employees, or artists actually need to have a completely stressed-out mind and totally scattered focus to perform their best, and this is why they must be yelled at, play in fear, feel a great deal of anxiety, suffer from constant threats, and be driven past the limits of their bodies' capacity. When you read psychiatrist Daniel Siegel's work on mindfulness—that he pursues as a doctor but also as a practitioner—he reminds you that mindfulness is as ancient as praying. In many different religious traditions from Christianity to Taoist teaching, you find the encouragement to "pause and participate in an intentional process of connecting with a state of mind or entity outside the day-to-day way of being."[27]

If we are willing to spend years and years learning baseball, gymnastics, math, finance, leadership, science, languages, music, computer programming, and so on, we need to find time to learn life-changing skills such as intention, visualization, and mindfulness. Besides, along with your mind and body performance, research consistently shows that mindfulness enhances intellectual outcomes.[28] If mindfulness practice is taught early, it becomes a default neural network in your brain, but if you missed out on it as a kid, you can start at any time in your life. Mumford likens the process to learning to ride a bike since they are both experiential learning practices. When learning to ride a bike, we fall over; we lose balance; we tip and then slowly but surely we ride; then we ride without thinking, without the mind intervening and getting in the way of our body's deep knowing.[29] Mindfulness is the same. It's experiential. You can't just read about it; you need to practice it, day in and day out. It takes time. You may feel, like I certainly did, that you are hopeless, that the whirling dervishes in your brain will never calm down. You will never enter into the present. You are doomed to the most random bits of song, fragmented memories, lists of things to do, thoughts about how your body is uncomfortable, and so on. I promise you that if you stick with mindfulness, experience it, practice it, slowly but surely your brain will slow down, clear itself of chattering gibberish, and move into the eye of the hurricane where you become aware, purposeful, curious, kind, and profoundly calm and present.

The bullying and abuse paradigm insists that you are a nobody. The only thing you are supposed to listen to are the commands of those in power. In fact, you should listen to their voices even when they hurl put-downs and humiliating slurs at you. There's no space for you to listen to yourself. They say jump and your only response should be "How high?" The bullying paradigm has trainers and coaches, like the ones at my son's high school or at the University of Maryland, who can effectively sever the deep bond between Mind-Brain-Body in a student-athlete like Jordan McNair. To remind you, Jordan McNair, along with his teammates, was repeatedly humiliated, degraded, and demeaned with homophobic slurs by their coach and trainer. The voice of the Mind-Bully dominated in this toxic culture. Despite his brain sending out alarm signals that his body was in danger and was entering into a heat-induced crisis during football practice, the Mind-Bully kept him trying to push through until he collapsed. Even then, committed to the bullying code, McNair's coaches and trainers did not get him the medical care he needed. He died two weeks later of heat exhaustion. He was nineteen.

These kinds of coaches and trainers are the opposite of a trainer and coach like George Mumford who operates in the new neuroparadigm. Mumford advises athletes to take time to *listen* to their bodies: "If you have pain in your body, don't struggle to get past it. Stop and listen to your body; surrender to being with what is. Ask yourself, 'What is the lesson for me to learn here?' 'Have I been overdoing it?' 'Have I been going too hard or too long?'"[30] Instead of internalizing the Mind-Bully, you have the choice to internalize an Empathic-Coach who, as we saw, Daniel Coyle has discovered is a talent-whisperer. Mumford tells his athletes: "Your body is like a circuit breaker; injury is its way of protecting you and telling you to change something. Learn to listen and to trust that still, small voice inside, the voice of self-knowing. You may slow down in the short run but it will keep you far healthier and more active in the long run."[31] Take a moment to get acquainted with your own "still, small voice inside, the voice of self-knowing." If you've been or still are the victim of bullying and abuse, be careful. Opening yourself up to moments like this, listening like this, can result in the Mind-Bully coming in loud and obnoxious and harmful. Not a bad idea, if that's the case, to go still and quiet and listen for your "small voice inside" with a mental health professional by your side. Being with a pro who can create a safe space for you to go to this place is important if you've suffered trauma in the past.

As we unpack the difference between coaches (or parents, teachers, bosses, etc.) who *make* individuals and those who *break* them, compare George Mumford's advice about listening to the "still, small voice inside" to the homophobic slurs of the bullying and abusive coaches of my son or of Jordan McNair. These coaches, well-trained in the bullying and abuse paradigm, believe they must break youth down with the goal of building them up. That clever reversal should make your head spin along the lines of "Wait, how can breaking create building? Aren't those polar opposites?" In contrast, Mumford's mindful, empathic advice can be found at the heart of the new neuroparadigm. No one needs to be broken down. No coach needs to break their athletes. It's okay to *disobey* the directives of the bullying paradigm.

The neuroparadigm, grounded in research, is all about *making* individuals and growing talent, not crushing it. Jordan McNair tragically ended up with a coach and trainer who adhered to the bullying and abuse paradigm. Not only did he fail to fulfill his athletic potential, the abuse he suffered resulted in his death as a teenager. In contrast, one of the greatest quarterbacks of all time, and the oldest quarterback to continue achieving at the highest level of his sport, has a trainer whose goal is to keep him in peak performance by strengthening his body and *his brain*. Tom Brady not only does daily workouts in the gym with his trainer Alex Guerrero; he also does the daily brain training designed by Merzenich and his team.[32] In the new neuroparadigm, brain training designed by neuroscientists is like mindfulness but more targeted. Tom Brady daily works on the brain "muscles" that help him make better split-second decisions, maintain excellent focus, connect socially and emotionally with his coach and teammates, draw on an exceptional working memory, and hone his peripheral vision. Following in his footsteps is another of the world's greatest athletes at this time: footballer Harry Kane has also publicly shared that he daily commits to Merzenich's brain training to excel.[33] Other elite athletes use the program but like to keep it discreet as they know it gives them a competitive edge. Isn't it tragic that teen athletes like my son and like Jordan McNair did not get to work with a trainer like Alex Guerrero, whose goal is to keep his clients' bodies safe, flexible, and enduring so that they can fulfill their potential? Instead, these teens, like many others, were coached in the bullying paradigm that hinders performance with humiliation and pain.

Doctor and pain specialist Michael Moskowitz defines chronic pain as "learned pain," and the steps outlined in *The Bullied Brain* are about *unlearning* thought patterns, destructive habits, and limiting beliefs, which are an awful lot like "learned pain" mechanisms that are no longer needed or useful.[34] In the process of repairing the harm done, fixing the "breaks," and becoming whole once again, you now need to *unlearn* the bullying patterns that may still reside in your brain. Like the bullying paradigm that produced them in the first place, these patterns are outdated. They no longer serve you. In fact, they may make you feel like you're being broken down rather than being built up. One way to unlearn these patterns is to supplant them with new ones. Here is Norman Doidge's description of exactly what Moskowitz did to oust his chronic pain maps.

> Each time he got an attack of pain, he immediately began visualizing. But what? He visualized the very brain maps he had drawn, to remind himself that the brain can really change, so he'd stay motivated. First he would visualize his picture of his brain in chronic pain—and observed how much the map in chronic pain had expanded neuroplastically. Then he would imagine the areas of firing shrinking, so that they looked like the brain when there was no pain.[35]

You can apply this same technique to whatever it is you want to change in your brain. It does not have to be chronic pain from an injury that healed. Repeated visualization works in a very direct way by using thought to stimulate neurons. In other words, the way you think or what you think about has the capacity to stimulate certain brain cells. "On brain scans," Doidge explains, "we can see signs of the blood rushing to the visual neurons of the brain that are being activated."[36] Try visualizing yourself first with whatever neurological scars you may have and any health issues that accompany them. Then consciously, purposefully, mindfully replace that map with one where you are unscarred and your health has been restored. Of course, because our experiences and brains are unique, this exercise could take many diverse forms. The takeaway is that you can acknowledge your pain and then you can use mindful visualization to calm the pain circuits in your brain.

Moskowitz is harnessing the unprecedented power of neuroplasticity. Just as Mumford has his athletes enter into a mindful state and then has them visualize athletic feats they want to achieve, the pain specialist recognizes an influx of pain, enters into a mindful state, and methodically visualizes a pain-free brain map. Note that he doesn't dismiss or deny the pain map. He acknowledges it and then sets about redrawing the picture in his brain until the firing up of pain is extinguished so that it cannot wire in. My son Angus's approach is different because his pain is triggered by anatomical rather than injury-generated pain. He has not healed like Moskowitz. In fact, specialists do not know how to heal the fused spine of Klippel-Feil Syndrome or the challenging and life-threatening conditions that accompany it. So while Angus cannot eliminate his pain, like Moskowitz does, he can use mindfulness to move himself from a darkened room subsumed with pain into a place he chooses that is safer, brighter, and full of wonderful memories and communion with friends and mentors. This empowers him and gives him space and time to hope and pray the medical community does indeed find a way to cure his pain.

Angus's honesty with me, when he said I could not handle the pain he daily endures, is an ideal reminder that strengthening desired neural networks is not a quick fix. It's not easy. It requires belief in yourself, followed by daily practice. It's as difficult as doing Merzenich's brain training, getting into top physical shape, and daily committing to mindfulness practice. Twice a week, Angus attends specialized physio sessions with his trainer Briana May at Neuromotion. Try to imagine having the worst migraine of your life and then going to an hour-long workout. It takes immense belief in himself and discipline to do it. He has to try to overcome his severe pain to maintain a baseline of health. Moskowitz describes the effort required by those who live with chronic pain: "I had to be relentless—even more relentless than the pain signal itself." That's the key term to remember, *relentless*. Changing neural networks or brain maps is not easy. It takes a massive amount of energy, effort, hard work, and belief that it's worth it and that you can do it. Moskowitz explains that every single time he felt a twinge of pain beginning, he'd have to close his eyes and visualize the change to "disconnect the network, shrink the map."[37] You too can disconnect the associations that bullying and abuse may have linked in your brain. Just like Moskowitz changed his brain and dissociated movement from chronic pain, you too can change your neural networks and brain maps. But you have to be *relentless*. You have to go at it like a warrior. Even though you may have been wounded, even though you may suffer from daily agony, you still have to believe in yourself because it takes time and effort to commit to deliberate practice.

There is a new technique being used successfully with actual warriors. Named "Virtual Iraq," the program is designed for veterans who suffer from PTSD. Veterans return to civilian life but find their war trauma perpetually triggered by sights, smells, sounds, and so on. A successful technique has been to have them replay the trauma with video-game-like software. They work very hard to reexperience the triggers of their trauma—a loud sound, a certain light, a particular movement—while mindfully disassociating it from debilitating reactions.[38] It does not happen overnight. It takes a great deal of belief in the process, then deliberate practice, along with empathic coaching. As with all other skill development, the more the veterans fire up the neurons for remaining calm and collected despite a trauma trigger, the more they wire together those calm and collected neural networks. Myelin (the fatty white insulator) wraps around the new neural network making it so efficient it becomes a default neural network. Essentially, the premise is that the veterans need to "return" to the site of their trauma, via video game, so that they can reprogram their brains now that they are safe. You can use this same technique with bullying or abuse trauma you may have suffered. Not a bad idea to do this kind of exercise with a mental health professional as of course it can be triggering and you want to ensure you are fully supported. Once you separate yourself fully from the bullying and abuse you may have endured, once you name the aggressors to break their hold over you, your brain becomes hungry for learning. Just as the body loves to move, the brain loves to learn. Never forget you are a warrior and should treat yourself that way. Since 2018, every soldier, sailor, airman, and marine has had free access to Merzenich's brain training, covered by the U.S. Armed Forces, because it supports peak performance, crucial in a life-and-death arena, but also because the training can repair traumas to the brain from explosions and from the psychological traumas experienced in war zones.[39]

One of Moskowitz's patients expressed what her life was like while her brain was ruled by chronic pain: "I was depressed and suicidal. And it didn't matter what drug the doctors gave me—the pain never went away. I couldn't even watch TV or read because, on top of the pain, the drugs I took put me in a gray zone. There was no reason to live."[40] Remember, emotional pain shares the same pain circuits as physical ones, and this description resonates with how I felt after Ellen killed herself and the whole crushing weight of the bullying paradigm came down upon me making me feel absolutely hopeless that change was possible. This is why when I read Moskowitz's approach, harnessing his neuroplastic brain to recover from pain, I felt hopeful. Step by slow, deliberate step, I believed that I could write a book that would protect others from feeling like Ellen had. With each page I wrote, each attempt to expose the falsehoods and manipulations of the bullying paradigm, I got that much farther out of the cage of learned helplessness.

It might have been foolish for me to wish to take on Angus's pain, as it requires beyond warrior strength and I have not earned that level, but all of us can start with small steps that will, slowly but surely, lead us to heal our pain whether it's caused by physical or emotional suffering. It took Moskowitz weeks of mindful visualization, which became months, which became a year. It required discipline and commitment, but here's the good news: he no longer suffers from chronic pain. His brain is no longer sculpted to overreact to every stimulus with a panicky shot of pain. Moskowitz used his neuroplasticity to calm down the pain circuits in his brain. You might say that he became responsive to stimulus, not reactive; in other words, he mastered mindfulness. He brought his brain into the present so that it doesn't send off alarm bells at every turn or every perceived threat. He lives pain-free. To remind you, the key neuroscience phrase that applies to this process is "use it or lose it." Because Moskowitz was not firing up, or "using," his pain maps, they faded away. Practicing "use it or lose it," he not only lost them; he drew new maps.[41]

While Angus, not suffering from injury but an anatomical abnormality, a spinal fusion, has not eliminated his chronic pain yet, he has developed a mindfulness practice that allows him to enter into the wilderness, walk, run, jump, travel, have intense connections with others, share triumph and tragedy with his tribe, battle for what's right, and have remarkable adventures without ever leaving his darkened room or removing his noise-canceling headphones. His mindfulness practice not only cuts through his pain, but it has opened a world within full of storytelling and fiction writing.

Teachers who now read his work do not say he's "pathetic" or "not trying." They say he's "gifted." They say they've "never seen a person so young write with such psychological insight." They say "his vocabulary is unbelievable." They say "the way he blends fast-paced, page-turning action with haunting descriptions is truly gripping."

The next step in Angus's pain journey, and in the pain journey of anyone who suffers emotional pain from bullying and abuse, is to intensify mindfulness with brain training. Merzenich advises that for someone like Angus, a strategy worth trying is to focus on the BrainHQ exercises in the "Speed and Attention" category. He explains that "speed training in vision or hearing drives physical and functional changes that are generalized to other systems via processes in the cerebellum." Whether working with human patients or in animal models, scientists document positive impacts in the "somatosensory brain." If an individual—with a brain suffering from physical or emotional pain—commits to brain training to improve processing speed or fast-timing skills in listening, the pain is lessened and the improvements felt can be seen as associated "neurological changes." In terms of pain reduction, whether from a spinal fusion or from bullying and abuse, Merzenich resorts to capitals, which is a good sign: "That generalization SHOULD have positive benefits for re-establishing the dominance of NON-PAIN over pain from the body region that hurts. It can't hurt Angus to try this—and it could HELP a lot."[42]

As a parent of a child with debilitating chronic illness, it's startling to learn that there is a way via neuroscience to lessen the pain. As a reader and writer about individuals suffering from extreme psychological pain due to bullying and abuse that is ruining their health, it's inspiring to discover that there is a scientifically established method for rewriting pain maps and replacing them with "NON-PAIN" maps. As a teacher and protector of a student who was abused and revictimized despite her fragile mental state, it breaks my heart that I did not know in time that "mental rehearsal drives brain plasticity" just as our actions do. I did not understand that "when we rehearse our misery, we embed it and it can grow." I wish more than you can know that I was able to tell Ellen that "it is easy to grow and sustain pain in the brain." It's dangerous and life-threatening when we are not informed just how much pain can overtake our brains. With chronic pain sufferers and with abuse victims, pain "commonly survives when the original source of the pain has evaporated."[43] But in our outdated and limited bullying paradigm, we believe when the body is healed, when the abusive individual is removed, then the victim is healed, safe, and pain-free. It is long overdue for us to debunk that harmful myth. In our brains, pain can linger and do devastating damage.

# Step 8: Redraw Your Brain Maps

As noted, you can exponentially intensify mindfulness practices by doing targeted brain training. If you want to begin with mindfulness as a way to prepare for the more focused brain training, Moskowitz offers a step-by-step approach worth trying. He has provided a method to teach you how to become a brain cartographer who erases destructive maps and redraws them to restore your health and happiness. He uses the acronym "mirror" to remind you of the six steps to easing or even erasing your pain.

- **Motivation:** You must actively visualize. The constant message of the bullying paradigm is that you are not worth it. You don't deserve it. It's your fault. You should be ashamed of yourself. Just voicing the desire to be whole and healthy might be enough to trigger the Mind-Bully, so be prepared and ready to assert your right to heal your scars and restore your health. Bring in or create your own Empathic-Coach, a talent-whisperer who believes in you unconditionally and with great compassion.
- **Intention:** You must focus on harnessing your mind to change your brain. Your goal can't be just to change your beliefs or habits because that sets you up for failure. You might struggle and make mistakes. You might fail some days. You might take two steps forward and one step back. That's okay and does not change your intention. The brain learns by making mistakes, so let it learn and stay crystal clear on your intention to heal your scars and restore your health.
- **Relentlessness:** You must be relentless like a warrior fighting against the odds. As soon as you feel pain, shame, associations with abuse, a sense of worthlessness, a sense of helplessness and despair, you must push back with everything you've got. And what you've got is unprecedented neuroplasticity so draw on

it. Neuroscientific research shows only intense focus succeeds to change brain maps. Avoid distractions to ensure you stay laser-focused.

- **Reliability:** The pain system evolved to protect: it's not an enemy; it's your protector. So you need to trust its signals and work with it. The emotional and physical pain you may have felt in the bullying paradigm wired into your brain because evolution was ensuring you remembered the threat and the danger and steered clear of it. So honor your pain, your protectors, and then let them go with the power of your mind to remap how you see and experience your world.

- **Opportunity:** Each hit of pain that you see as a gift, especially within the context of being able to ultimately stop it, has the capacity to change your mindset and thus your brain chemistry. You may have been "electrocuted" in the past and suffered intensely because of it, but it's over now, the door is open, your Mind-Brain-Body are a powerful triad that can march you right out of the cage of learned helplessness. You can unlearn limiting beliefs and destructive habits. You can replace them with limitless potential and healthy choices.

- **Restoration:** To heal your scars and restore your health, you need to be responsive, not reactive. Mindfulness practice strengthens the neural networks that allow you to respond to situations and to others without getting triggered and stressed. Approach your pain this way. You are not fighting the pain; you are not fleeing the pain; you are not freezing faced with the pain. No. You are restoring healthy, whole, integrated brain function by activating at will your parasympathetic nervous system. The deep, calming breaths and the visualization communicate from your mind to your brain and body that you are safe, protected, calm, and in power.

Moskowitz did not only succeed in removing his own chronic pain; he is also successful with many other sufferers after he teaches them to gaze into and then use his mirror.

# CHAPTER 9

# Restore Your Brain with Your Body

## Step 9: Oxygenate Your Brain

In his final year of high school, Montgomery spent most of his time alone in his room, lying on his bed, staring at the ceiling. I had never seen him immobile before. He was a born athlete, a kinesthetic learner; he excelled at any sport he tried. We have pictures of him as a toddler with his plastic hockey stick, blocking shots in the doorway to our kitchen in Toronto. The baseboards ended up being scuffed with black marks from the puck. He started skating at two, began hockey at five, and continued to play competitive ice hockey while also playing soccer and rugby for a year. While he was asked to play on a rep hockey team and a grade level higher in rugby, he decided to step down from both to concentrate on basketball. He tried rowing for a year and won gold medals with his quad. He loves to play squash, tennis, golf and jumps at the chance to go hiking or skiing whenever he can. But basketball has always been his true love. He would shoot for hours on the basketball hoop affixed to our garage. Playing on his first team when he was eight years old, he slowly but surely over his teen years let the other sports go so he could dedicate himself to basketball. Even when he was walking, he would always be taking phantom shots and dekeing out invisible players.

Now he just lay on his bed wasting away.

In April, the year before, the headmaster encouraged students to give testimonies about the abuse they had suffered on the basketball team. With each passing day the crisis expanded. It wasn't just the Senior Boys team that was reporting abusive coaches; it was also the Senior Girls team and then the Junior Boys team. The more the abuse situation grew, the more contradictory directives were being issued from the headmaster and board. The headmaster wrote that he'd spoken with the Senior Boys coaches and they were sorry, especially about how they had treated Montgomery. They said they didn't know how much harm they'd caused. The next day, the headmaster wrote to say he had to believe they were "innocent." It was confusing. He promised there would be different coaching, and he assured the students—who gave detailed testimonies about the abuse—that they would be protected with confidentiality. He did not keep either promise.

Montgomery had wanted to apply to college basketball teams as the sport was his absolute passion and he had the talent. However, he refused to play with the two abusive teachers put back in the position as coaches. Six boys refused to play that year. While it was a healthy decision from a Mind-Brain-Body perspective, he nonetheless slipped into a severe depression. Not playing the sport he loved, a sport that was so important to his sense of self and his future, being revictimized by the headmaster and bullied for speaking up, it was all too much. He went from being on the headmaster's honor roll to missing 172 days of school, not writing his end-of-year exams, and not even attending his graduation ceremony. Far worse than all that, he couldn't play with his team. Normally, he'd be practicing and playing basketball games six days a week, and even under abusive conditions, it seemed less traumatizing than having his sport taken from him. Regardless of a rigorous basketball schedule that started in September and went through April, the second Montgomery was on holiday, he'd go to basketball camps. He couldn't get enough of the game. Now he was in free fall. He played soccer and continued his outdoor education leadership program, but it wasn't the same. He played golf and tennis with his father and friends, but nothing could replace competitive basketball. Most days, he lay with his eyes closed or stared at the wall. He spoke little.

A victim of abuse at the hands of high school teachers, I had suffered a severe depression myself in my final year so I knew how hopeless and trapped Montgomery must feel, but I did not know how to help him. He had weekly sessions with an excellent sport psychologist, but while he learned how much he had being playing his sport in a state of extreme fear in previous years, he did not seem able to shake the despair at what had been taken from him by the very people who were supposed to support his learning and success. This betrayal had pushed him into a very dark place.

And then one day in the spring, after months of isolation and misery, I saw him lace on his running shoes. He walked out the front door without looking back.

From that day forward, he began running for half an hour to an hour a day.

Montgomery doesn't jog. He runs. He's six four and a half and 172 pounds of muscle. I'd watch him from the living room window at the tail end of his run when he needed to go up a major hill to reach our house. After a month, he was floating up the steepness as if he was descending.

After university, Montgomery moved to Vancouver, British Columbia, for work. There's a mountain at the edge of the city, called "Grouse," and it has an ascent that is steeper than stairs through the rocks and trees. It takes a fit person probably an hour or so to reach the top and then take the gondola down. As often as he can, Montgomery runs up the "Grouse Grind" as if chased by demons, leaving the other athletes in his wake.

While I didn't know then, I've learned now that aerobic exercise is one of the best methods to heal scars in the brain and restore the health of the Mind-Brain-Body. "In 2010, the American Psychiatric Association issued new guidelines for treating depression, and for the first time, exercise was listed as a proven treatment."[1] Montgomery was lucky because he is drawn naturally to movement and sport, but the rest of us also can tap into the incredible power of our bodily selves to recover from traumas done to our brains.

One way to get a sense of how the bullying paradigm fragments us is by considering for a moment the way it frames our brain and body as separate entities, strangers even. Operating within the framework of what Antonio Damasio calls "Descartes' Error," as we discussed before, schools believe they must limit, or even remove, recess and physical education classes. Why do they do it? Because they believe that it improves students' academics by providing more time for learning and studying. This makes zero sense from a brain perspective. In 2007, the American Academy of Pediatrics (AAP) set out to do research on recess. Their hypothesis was that recess was important for children from a physical standpoint. They were surprised to discover that in fact, recess benefited the whole child. The social, emotional, cognitive child greatly benefited from movement and play.[2] Even though this research is now well-known and has the stamp of approval from the highly respected AAP, there are still those adherents of the divided child, framed by the bullying paradigm, who want to do away with recess and running around, playing outside, all in the name of sitting kids at desks, ensuring they obey, and pouring information into them.

While we turn to physical activity as a way to halt the obesity epidemic in youth populations, few realize that physical activity and play are equally as vital for brain health. It is well-documented that a sedentary lifestyle is correlated with "lower cognitive skills."[3] If we have concerns about declining literacy and numeracy standards, we would be wise to ensure that, as part of a health plan, *all* children are able to play actively and physically at recess, in physical education class, and in sports or other forms of exercise. There is still a widespread belief that making children study academics all day in a classroom, at a computer, or poring over books is going to improve their intellectual capacity. The fact that after a day of school, children are sent home with hours of homework to complete indicates that educational authorities do not understand the immense brain benefits of exercise and free play. As we saw in the Adverse Childhood

Experiences study, obesity is not primarily about a lack of physical exercise. People eat as a way to protect themselves against sexual violation or physical violence. Merzenich reminds me that "eating is controlled by the brain." His colleague and friend neuroendocrinologist Mary Dallman showed in her research that "stress leads directly to a craving for carbohydrates." Many of us believe that if we simply included exercise in our lives, we would lose weight and be healthier. While this is certainly true, Merzenich stresses that, as with everything else, we need to start with the brain and address any "neurological distortions" that may drive excessive or unhealthy food consumption.[4]

Molecular biologist John Medina argues for the importance of exercise: "Cutting off physical exercise—the very activity most likely to promote cognitive performance—to do better on a test score is like trying to gain weight by starving yourself."[5] Merzenich notes that "controlled trials where individuals exercise" just their bodies, just their brains, or both provide insightful and applicable findings. If you want to improve cognitive performance, the research shows that "direct brain exercise is substantially more powerful." As we discussed in chapter 6, your brain needs and craves workouts as much, if not more, than your body. Keeping your brain "fit" is where the research finds the most significant strengthening of cognitive performance. Add in physical exercise and you have an even greater capacity for a healthy, strong, high-performing brain. If you only exercise the body, your results drop considerably.[6] The key is working Mind-Brain-Body as a holistic trio for optimizing performance in life.

While some contemporary educational legislators today appear out of touch with twenty-first-century brain research, *you* don't have to be. If you attended a school that cut recess, trapped you in a desk for too long, drove you away from enjoying PE class or after-school sports with a bullying coach, or medicated you when you couldn't concentrate due to lack of exercise, then it's time to change your mindset. There aren't any "breaks" within your Mind-Brain-Body. They are inextricably entwined into a beautiful whole. If they've been sundered by the outdated bullying paradigm, now is the time to reject its fragmented approach and heal the damage done. There is "a massive pile of evidence that says the quickest, surest path to the health and well-being of the brain and body is movement, or vigorous aerobic exercise."[7] Note "brain and body" together, not separated. This echoes the holistic model of medicine researched by Felitti and Anda; it echoes Damasio's linking of cognition and emotion in learning. Psychiatrist Stanley Greenspan stresses the whole self as well. "The basic element of thinking—the true heart of the creativity central to human life—requires lived experience, which is sensation filtered by an emotional structure that allows us to understand both what comes through the senses and what we feel and think about it as well as what we might do about it."[8] If you have had lived experience that is adverse, you may well have built an "emotional structure" within you that has strong negativity bias; is riddled with anxiety; chooses passivity as it feels safer; feels weighted with depression, hypervigilance, or aggression; and so on. What's encouraging about aerobic exercise is that it has the capacity to topple the limitations of lived experience and open us up to creative new ways of experiencing it on a brain level.

Merzenich refines our understanding of exercise and its role in keeping our brains strong and healthy. You need to distinguish between forms of exercise and their respective impact on your brain. "Highly repetitive exercise" such as running on a treadmill or using a rowing machine "has less value than progressive exercise forms that grow your fundamental neurological control of your actions." Examples of progressive exercise would be learning soccer (or any sport), running or biking off pavement, kayaking in white water, skiing, dancing, yoga, martial arts, and so on. Just think about how hard your brain has to work in these kinds of physical endeavors. This kind of exercise "GROWS your executive powers" because it engages "the same controlling machinery that controls your operations in thought."[9]

Neuroscientist Norman Doidge discusses research done by Gerd Kempermann, who is a multiple-award-winning professor for "Genomics of Regeneration" at the University of Dresden. Kempermann studies mice who have environments that allow them to actively *play* with "balls, tubes, and running wheels," while other mice in the experiment are in environments that lack these activity-promoting toys. When Kempermann examined the brains of the mice who played for forty-five days, he discovered that they had a "15 percent increase in the volume of their hippocampi and forty thousand new neurons, also a 15 percent increase, compared with mice raised in standard cages."[10] In other words, if you get moving and playing, you can gain forty thousand new brain cells. If you don't get active, you don't. This increase in volume in the hippocampus (remember, it's the area in your brain involved in learning, storing memories, etc.) went up fivefold when the mice were given up to ten months in the enriched cages that encouraged active play. While the mice had healthier bodies, they also greatly outperformed the other mice in "tests of learning, exploration, movement, and other measures of mouse intelligence." Doidge concludes: "Physical exercise and learning work in complementary ways: the first to make new stem cells, the second to prolong their survival."[11] Merzenich points out that while exercise is key, it's important to see the larger aspects of the environmental change. One set of mice were living in a "featureless cage in which they were literally bored to death." These mice who were isolated and did not have opportunities to work their brains or bodies had shortened life spans. He compared the bored, lonely mice to mice who suddenly found themselves in a social toy- and challenge-filled environment. They could play. They could run. They were part of a community. Merzenich notes that it was the "mastering" of that new environment that is "expressed in strong form in their brain" and that we must factor in much "more than physical exercise." Once again we see the need to work with a holistic self, not a fragmented one, not one that is separated into silos.

In *Spark: The Revolutionary New Science of Exercise and the Brain*, psychiatrist John Ratey outlines the remarkable benefits that you gain from playing sports or participating in aerobic exercise that moves your body and increases your heart rate. Extensive research shows physical activity leads to biological changes that "encourage brain cells to bind to one another." Getting your heart rate up provides an exceptional stimulus that opens your brain up to learning, thinking, remembering, and problem solving. Aerobic movement impacts learning directly, at the cellular level, improving the brain's potential to log in and process new information.[12]

Exercise not only enhances physical health (your body) and intellectual ability (your brain); it also reduces stress, which is healthy and restorative for your holistic Mind-Brain-Body. Regular aerobic activity is a workout for your *brain* that trains it to better handle stress: "In the brain, the mild stress of exercise fortifies the infrastructure of our nerve cells by activating genes to produce certain proteins that protect the cells against damage and disease. So it also raises our neurons' stress threshold."[13] Exercise not only makes your muscles more resilient through repeat doses of stress; it does the same thing for your brain's "infrastructure" and your "neurons" (namely, brain cells). When sweat is rolling down your face, your lungs are burning, your legs are aching, take a moment to celebrate how these signs of stressing your body—to make it more adaptable, resilient, and powerful—are also happening to your brain.[14] Just because you can't see it or feel it doesn't mean it isn't happening. Remember the increase in hippocampal volume in the playful mice? Ratey explains that growth factors such as FGF-2 and VEGF are *both* produced in your brain and generated by muscle contractions, after which they "travel through the blood stream and into the brain."[15] Key word is *both*. Not in one or the other, as if you are a divided, fragmented, or broken being. No, in *both* your brain and body. Why? Because they are part of a complex, amazing whole.

To further cast doubt upon the premise of the bullying paradigm that we are divided and our brain and body do not have an inextricable influence on one another, let's look at the benefits of aerobic exercise (moving your body) on your brain. What does aerobic movement do to enhance healthy development in brains? Ratey replies that it:

- "kick starts the cellular recovery process";
- "increases the efficiency of intercellular energy production";
- helps "prevent the onset of cancer and neurodegeneration";

- "increases IGF-1, which helps manage glucose levels" and "IGF-1 increases LTP, neuroplasticity, and neurogenesis"; and
- "produces FGF-2 and VEGF, which build new capillaries and expand the vascular system in the brain. More and bigger highways means more efficient blood flow."[16]

When you work out or jog or play sports, it's amazing to think that as you become stronger, leaner, more resilient, and more flexible in your body, at the same time, your brain is in a state of recovery; it's preventing disease; it's experiencing more plasticity and the generation of new cells; and it's laying down more efficient pathways for oxygenated blood. It's nothing short of remarkable.

> Movement places demands on the brain, just as it does a muscle, and so the brain releases BDNF, which triggers the growth of cells to meet the increased mental demands of movement. But BDNF floods throughout the brain, not just of the parts engaged in movement. Thus, the whole brain flourishes as a result of movement. It provides the environment that brain cells need to grow and function well.[17]

To remind you, BDNF stands for "brain-derived neurotropic factor," which Ratey describes as a fertilizer for the garden of the brain. While this healthy reaction, produced by movement, occurs in the body and brain as measured by scientists, how might it look in a school that has decided not to limit or eliminate recess and PE classes?

Let's take a look at a school district that has embraced and led the way into the new neuroparadigm where research prevails and it's understood that the more physically fit students are, the better they do at their academics. There are schools that are constructed on the understanding that the brain and the body form an inextricable whole and when you exercise one, you exercise the other. Ratey raves about school administrators and teachers in the school district of Naperville, Illinois, who teach fitness instead of sport. "They're getting kids hooked on moving instead of sitting in front of a television."[18] Further supporting Kempermann's research on how much brain development occurred for mice who were able to actively play, the students in this fitness program have marked improvements in their academic achievements.[19] At Naperville, only 3 percent of students are overweight, in contrast to 30 percent nationwide, and these students who are fit and at a healthy weight are also outranking their peers in academics. In fact, these fit students have become some of the "smartest in the nation." In 1999, eighth graders at Naperville were among the 230,000 students from around the world who took an international standards test, the TIMSS (Trends in International Mathematics and Science Study). In past years, students in Singapore, China, and Japan surpassed American students in the core subjects of math and science, but the eighth graders in Naperville were the notable exception. When these students took the TIMSS, they took the *number one spot* in science and finished sixth in math.[20] The spectacular academic results at Naperville provide clear evidence of the correlation between students exercising (body) and students excelling at subjects such as science and math (brain). Neuroscientists using EEG scans discovered more activity in fit students' brains, which meant that more neurons engaged in attention were being recruited for a given task. This translates into more integrity or *wholeness*: better fitness results in better attention and better academic results.

Playing sports, practicing martial arts, and doing yoga or dance are activities that require challenging thinking as well as physical exertion. Anyone who has coached or played a sport understands in a profound way that physical movement combines aerobic activity with strategic thinking. Thus, athletics have been shown to "help promote the development of self-regulation and initiative," which are crucial life skills.[21] It only takes one hour per day to increase your "self-control," which in turn can "facilitate learning and achievement."[22] Ask your Mind-Brain-Body: Do you want me to oust my inactivity and inertia, my sense that I'm trapped? Do you want me to replace destructive habits with ones that heal my scars and restore my health? If they answer *yes* in unison, indicative of their wholeness, then you have *got* to find one hour per day to invest in aerobic activity, boosting the self-control needed to get rid of harmful habits and ramping up your learning and achievement.

Fitness is an "indispensable tool" for you to reach your full or *whole* potential.[23] Fitness is not only pivotal for your physical health but also for your mental health. Ratey notes that in contemporary life, "people tend to have fewer friends and less support, because there's no tribe." The bullying paradigm doesn't only strive to divide and conquer your own Mind-Brain-Body; it also isolates people, pits them against one another, unfairly privileges some while castigating others. And yet being "alone is not good for the brain."[24]

While seven out of ten kids quit sports these days at the age of thirteen (right when adolescent brain development makes them vulnerable to criticism) and while there's an obesity crisis reaching epidemic proportions in youth populations, there are also amazing programs like Naperville that simply sidestep the whole issue by focusing on the holistic Mind-Brain-Body, privileging it rather than the games. In other words, the children and the health of their Mind-Brain-Body are the focus, not the games they play.

When it was learned that concussions can actually destroy brains, many rose up in a fury to say that no one, and I mean no one, not even neuroscientists or parents, could dare to interfere with the rules of the game. If kids' brains were being damaged, well then, that's the price they paid for the privilege of playing football, hockey, soccer, or rugby. Parents who tried to protect their kids were seen as pathetic. Angry individuals jumped up on soapboxes and deplored the "wussification" of today's youth.[25] In the bullying paradigm, sport games and adult egos take precedence over children and their health. The battle that Kimberly Archie has had to wage to see children protected in sports and in cheerleading has been nothing short of disturbing.[26] Not only has she fought an uphill battle to hold sports organizations legally accountable,[27] she has been bullied relentlessly because she advocates for children's brain safety.[28] Indicative of the bullying paradigm's lack of empathy is that Archie's fight for child safety was sparked by her own son's death and her discovery that after years of playing football as a kid, he suffered from a traumatized brain. When I interviewed Archie, what stood out most was her ability to use the law to expose the bullying paradigm. Lawyers hire Archie to instruct them on the way that the law can support child safety and the rights of children to *not* be permanently destroyed by adult "games." Her motivation wasn't reputation or money. It was the death of her son; it was the injury of her daughter. She refused to believe anymore in an outdated, destructive paradigm that sacrifices children to adult egos or beliefs. Archie is not prone to melodrama, quiet even, but every word she speaks exposes the lie that children are safe in the bullying paradigm.

The school district in Naperville, Illinois, exited the bullying paradigm that argues sports come first and children and their rights are irrelevant; they are merely meant to serve adult needs. The school district in Naperville operates with the understanding that should structure all schools: "Sedentary behavior causes brain impairment, and we know how: by depriving your brain of the flood of neurochemistry that evolution developed to grow brains and keep them healthy."[29] Schools are believed to be places where brains are enhanced, not impaired. How is it possible that so many schools have children sitting and not moving for hours at a time? How can we be a society that drives seven out of ten children away from playing active games and then complains about them doing sedentary activities like video games for great stretches of time? Colleges and workplaces are mostly sedentary, and yet we believe that we must remain inactive to produce cognitively when the research says we could produce more and better from our brains if we were moving. If we didn't have the knowledge, we could excuse ourselves, but we have a massive amount of evidence that our brains best learn when our bodies are active.

Naperville wanted to see if offering children fitness opportunities, and lots of them, instead of having them do sports chosen by adults, made more kids want to remain active. And it certainly did. This is not to say that sports aren't wonderful because they are, but they don't appeal to all children and some practices put the brain in danger. If we want healthy children who optimize Mind-Body-Brain performance and become "athletes-for-life," then we need to widen our understanding instead of having a narrow view of exercise as dependent on traditional sports and adult egos.

The essence of physical education in Naperville is that they teach *fitness* instead of sports. The underlying philosophy is that if physical education class can be used to instruct kids in how to monitor and maintain their own health and fitness, then the lessons they learn will serve them for life.[30] Imagine teaching for life, not for the test or the grade or the quota or the outdated belief system. Instead of getting boys to try and do chin-ups that 60 percent can't do and getting girls to run laps and then lambasting them for not being strong enough or "looking slow," the teachers at Naperville create opportunities in their fitness program for success. Students are graded on how much effort they put in, and the teachers measure their fitness levels with heart monitors, not judgments or put-downs. This was transformative for the teachers, who realized that fast or athletic students did *not* have better fitness outcomes than a student exercising at their own pace in their own way. Remember, bodies and brains are unique; they can't be measured accurately against one another. Some students who "looked slow" turned out to be hitting the same heart rate seen in elite athletes. That's why some teachers at Naperville feel regret about turning kids off exercise in the past because they made them feel weak or slow and ashamed about it. Now the teachers have set their sights on the future, a new neuroparadigm. They make sure that they are training children to become athletes-for-life.[31] In other words, the teachers exited their own past training and belief system, hammered into them by the bullying paradigm, and they entered a *whole* new way of thinking, teaching, and understanding.

A study at Virginia Tech revealed that schools that reduced time spent in gym class to improve academics—namely, math, science, and reading—*failed* to improve their students' test scores. This is what happens when you believe in the outdated bullying paradigm that has no research to back it up. It's like believing that global warming isn't happening when you can look on a thermometer and see that it is in fact happening. If you're still not convinced, you can check out five years of data collected by the California Department of Education, which has consistently shown that students who have higher fitness scores also have higher academic scores.[32] Moreover, studies show that there is a marked drop in incidents of school violence when PE is every day, not just once a week.[33]

An important takeaway from the Naperville program is to remember your brain and body are unique and so you need to find the physical activity that suits you, that actually makes you happy. Sport physiologist Craig Broeder notes the critically important need to give individuals choice when they are doing fitness training. He stresses that you need to be successful and happy doing exercise or you will simply not continue doing it. He notes how "old school" PE classes limit student opportunities to one sport and then teach it like a drill sergeant at a boot camp. The drill-sergeant approach may well be a turn-off for a student who might be drawn to a whole range of different ways to become fit. In contrast, Broeder explains that at "Naperville, they give kids lots of options by which to excel; they design lifetime fitness activities."[34] He documents the shift from the boot camp mindset to a new approach whereby Naperville teachers see themselves as the "sculptor of bodies, brains and minds."[35] With the knowledge we have from neuroscience, we can exit the "old school" bullying paradigm and enter a "new school" *neuroparadigm*. You can find an Empathic-Coach or become your own "sculptor" of your Mind-Brain-Body. You can sculpt your body, turning it into a fit, strong, flexible, and resilient form, just as, simultaneously, you can sculpt your brain, inside your skull, to be a fit, strong, flexible, and resilient organ.

Naperville's exceptionality illustrates just how much aerobic exercise, chosen by students, with empathic coaches and teachers, leads to healthier, more focused, happier students who go on, not surprisingly, to shine in their academics. However, as Merzenich points out, and it is a critical point, the median family income in Naperville is $118,000 per year; the average across Illinois is $69,000, and in a city such as East Saint Louis, Illinois, it's $20,000. Gym classes are key, but there are other factors that fuel Naperville's success. If physical fitness was the marker for personal and professional success and all-around brain health, then we could expect athletes and physically fit military personnel to be our most advanced citizens in terms of cognition and emotional regulation. However, that is glaringly not the case.[36] As noted, we need to factor in traumas and then take a three-pronged approach, with evidence-based brain training arguably the most important intervention. We are not just a body, not just a mind, and not just a brain. Therefore, we need to align these three powerhouses to recover from adversity and to optimize our potential.

Exercise that includes strategy, teamwork, and skill development that constantly surprises and challenges the brain is going to take us much farther than simply running on a treadmill. Likewise, mindful exercise supports the self-regulation and self-discipline, as Laurence Steinberg explained, that also emerge with mindfulness practice. It's not by accident that neuroscientists like Daniel Siegel and Rick Hanson also are meditation teachers.

As we discussed in the last chapter, we can harness mindfulness to help alleviate stress in our lives, and it's the same with exercise. As we saw in chapter 6, these practices can effectively lay the groundwork for brain training. Researchers have discovered that unlike all other animals, our brains are so powerful that we can get stressed-out just *thinking* about a threatening situation. We don't even have to be in any kind of real danger. As neuroscientist Stan Rodski explains, when we merely imagine or think about a threat, our sympathetic nervous system raises the corrosive stress hormone in our brains, and we essentially think ourselves into a frenzy.[37] That's the bad news. The good news is we have *choice*, which means we can also "literally run ourselves *out* of that frenzy."[38] We can lace up our running shoes and go for a mountain jog; hone our tennis or ping-pong game with a partner; work on our passing in soccer or dribbling in basket-ball; practice synchronized swimming, ballroom dancing, or figure skating; do tai chi, kung fu, or Sistema; or ski, ride horses, throw balls, shoot hoops, skate-board, surf, or hike. Point being, aerobic movement proactively calms down our sympathetic nervous system, lowers the cortisol, and chills us right out. Next time you go for a run (or do whatever aerobic activity it is you want), do a scan of your stress levels before and after. Neuroscientists can guarantee you will see a marked difference. If you're stressing out, imagining the worst, riddled with fear and anxiety, all you need to do is put on your skates, riding boots, skies, cleats, flippers, or hiking boots and use exercise to become present, focused, and calm. Life in the twenty-first century, still stuck in a bullying paradigm, is a stressful place. For many of us, stress does not arise and then fall away when we recognize we are safe; stress has become relentless and chronic.[39] Chronic stress, as we've discussed previously, is extremely harmful for your brain and body.

Stress in small doses is healthy and effective for your brain and body. Like a vaccine, small doses of stress that produce small doses of cortisol can help to wire in memories, which is great for learning, but it's always about the tipping point as "too much [stress] suppresses [memories]; and an overload can actually erode the connections between neurons and destroy memories."[40] Not only does this show why stress harms learning and exercise enhances it, but it is also a sharp reminder that victims may struggle to report bullying and abuse because their brain's reaction is to literally lose the traumatizing memories, which then can be triggered and return as inarticulate *fragments*, as explored extensively in the work and research of psychiatrist Bessel van der Kolk.

Starting an exercise program, brain training, or a mindfulness practice can feel like a daunting challenge. I find that tapping into flow makes it easier. I'm not minimizing how tough it is to get physically, neurologically, and mindfully fit, believe me. I know the painful, slow process of turning passivity into activity or lack of confidence into belief in oneself. It is incredibly hard to transform a fragmented Mind-Brain-Body that are at cross-purposes into a highly functional, integrated, symbiotic, holistic self. It takes many, many sessions of lumbering awkwardly down the street, but one day, that awkwardness becomes confident, competent jogging, which can even morph into the blissful moment that you experience runner's high and begin looking forward to that amazing feeling of endorphin release generated in the brain. It takes time and effort, but it is worth it. You can replace traumatizing experiences with joyful experiences of flow, where you forget the world around you. You forget time itself and simply become one with movement, cognition, and creativity.

As psychologist Mihaly Csikszentmihalyi reminds us, flow was at the original heart of the Olympics: "The Latin motto of the modern Olympic games—*Altius, citius, fortius*—is a good if incomplete summary of how the body can experience flow. It encompasses the rationale of all sports, which is to do something better than it has ever been done before. The purest form of athletics, and sports in general, is to break through the limitations of what the body can accomplish."[41] We should strive for *altius* (higher), *citius* (faster), and *fortius* (stronger) not just with our bodies but also with our brains. Let's take this one step farther and say: the purest form of athletics, and sports in general, is to break through the limitations (or fragments) of what the Mind-Brain-Body can accomplish by once again returning them to their proper state of integrity and wholeness. A flow state can be achieved when you simply lose yourself in the joy of exercise, cognition and creativity, or mindfulness. Csikszentmihalyi is concerned that many "people get caught up in a treadmill of physical activity over which they end up having little control, feeling duty bound to exercise but not having any fun doing it."[42] I'm sure you recognized the ominous seven words that are red flags of the outdated, bullying paradigm's way of thinking: "having little control" and "not having any fun." Those are also key phrases that show up in discussions of chronic stress. And those key phrases are exactly what Naperville schools strive to eliminate. They have kids experience exercise as something *they* have control over, as it is their choice what activity they want to do. They are able to have fun because their coaches see themselves not as drill sergeants but as sculptors of bodies, minds, and brains.

When it comes to the bullying paradigm, we are so indoctrinated that we forget that our brains have the "shapeshifting" capacity to change and to create. As neuroscientist David Eagleman reminds us: "The brain chronically adjusts itself to reflect its challenges and goals. It molds its resources to match the requirements of its circumstance. When it doesn't possess what it needs, it sculpts it."[43] The challenge is to exit the bullying paradigm and sculpt a new neuroparadigm, one that makes training the Mind-Brain-Body and creating healthy, holistic selves a topmost priority. That leaves no "cortical real estate" for bullying and abuse.[44] Eagleman quotes neuroscientist and Nobel laureate Santiago Ramón y Cajal, who, long before brain imaging was even dreamt of, said: "Every man can, if he so desires, become the sculptor of his own brain."[45]

Coach Phil Jackson articulates how blending exercise, strategy, honed skill, teamwork, and mindfulness unites Mind-Brain-Body and offers a powerful alternative to the falsehoods issued in the bullying paradigm.

> When players practice what is known as mindfulness—simply paying attention to what's actually happening—not only do they play better and win more, they also become more attuned with each other. And the joy they experience working in harmony is a powerful motivating force that comes from deep within, not from some frenzied coach pacing along the sidelines, shouting obscenities into the air.[46]

The word "obscenities" hits me like a punch in the face because it is the word used by the Commissioner for Teacher Regulation, who wrote that he would not hold the teachers accountable for their abuse because the students should not have "listened to the teacher's obscenities." In the Orwellian world of the bullying paradigm, students who report teachers' bullying are blamed for it because they listened to it! In the new *neuroparadigm*, a championship-winning coach like Phil Jackson harnesses the uniting force of mindfulness as his athletes play sports together to create harmony and overcome divisiveness. Mindfulness here has the force of empathically connecting and bonding players. There is attunement deep within that manifests as attunement with one another. This is the opposite of the fragmenting, dividing force of the bullying paradigm that hurls obscenities from the sidelines while claiming it is for the players' own good.

A fragmented Mind-Brain-Body clangs and jangles, whereas mindfulness in movement leads to attunement and harmony. The motivation is intrinsic, which leads to growth-mindset, not extrinsic, which leads to fixed-mindset. In other words, you have trust within your united Mind-Brain-Body as it works together in healthy, powerful ways and serves to connect you to others. Jackson sounds like he is describing a Csikszentmihalyian flow state in his players when they practice mindfulness in movement. Notice too that these athletes "play better and win more." The bullying paradigm loves to argue that if athletes aren't yelled at, sworn at, humiliated, and driven to the point of exhaustion, injury, and strain, then they won't perform. They'll lose. The bullying paradigm is constructed on the false belief that players need to be *broken* down and then built back up in an image that reflects the coach. To my mind, this is a powerful example of narcissism.

Montgomery's team had played together since fourth grade. They won every game up until tenth grade, when they lost in the provincial championship to another team. This was the first year they began being coached by one of the teachers identified as abusive. The following year, this abusive teacher was joined by another abusive teacher, eliminating any checks and balances from other coaches, and the team fell apart. The older boys were pitted against the younger ones. Some players who were given positions and privileges they hadn't earned were pitted against the players who were denied opportunities they merited. Some players were guided with an arm over the shoulder and a directive about what to do while others were berated and shamed. Other boys were outright ignored as if they didn't exist. It couldn't have been less harmonious, less mindful, or less successful. As noted, Montgomery was one of the targets. The coaches publicly recognized that he was one of the best, if not the best, player on the team and one of the best players ever in the history of a school a hundred years old.

Now, after completing his BA at the University of Oregon, Montgomery works in the Vancouver film industry, as a camera operator. In film, the whole production hinges on teamwork. As an empathic person, keyed into what his colleagues think, feel, and intend, he is very successful. Identifying his teachers' treatment as abusive, reporting to the headmaster, and finally speaking to national media ensured that Montgomery did not identify with the aggressor, and he does not perpetuate the bullying paradigm. Luckily for him, he's far more fit, far more of a team player, and far more attuned and mindful than he ever was when being abused. The motivating force that he experiences deep within himself has never, and will never, come from some frenzied coach pacing along the sidelines, shouting obscenities into the air. Unfortunately, for every youth like Montgomery who succeeds in "not letting them break me," there are countless fragmented victims who spend their lives trying to piece themselves back together.

# Step 9: Oxygenate Your Brain

When breathing, we tend to think about our lungs, but it helps—whether practicing mindfulness or doing aerobic exercise—to think about your brain. Your brain is so active, all the time, that even though it's only 2 percent of your body weight, it uses 20 to 25 percent of your oxygen. While you spend a good part of each twenty-four-hour cycle immobile while sleeping, your brain is continuing its activity, drawing on your oxygen.[47] On one level, our brains specialize in movement, and it's one of their defining features that cannot be replicated by Artificial Intelligence. AI can learn how to win a chess game, but it cannot learn how to move a chess piece from one square to another. British scientist Daniel Wolpert says "We have a brain for one reason only: to produce adaptable and complex movements. There is no other plausible explanation."[48] Merzenich exhorts us to take this to heart and get out into the world for our health. "Information from the body, senses, balance organ, and vision control the adjustments in the moment-by-moment flow of blood that must be accommodated as you change your posture. We very heavily exercise this vascular control machinery in a natural environment."[49] Exercise in a gym is "impoverished" from a brain perspective, and thus we must set forth and move in nature to activate our senses, our brain's excitement about surprises, and our in-born need to adapt and remain flexible. If we are filling our brains with oxygen, then it's best to draw on the fresh, oxygenated air from trees.

While neuroscience has used the analogy that the brain is *like* a muscle because new neurons and brain networks grow as needed, "new-era neuroscience says that the brain *is* a muscle."[50] This muscle in our skull not only needs exercise; it needs wild exercise, challenged by nature, our "ancestral home."[51] Our ancestral home is full of oxygen that our muscles and brain draw on when we engage in aerobic exercise. Ratey and Manning want us to understand that there is nothing more natural or beneficial than exiting the cage of our sedentary and civilized lives and moving in nature: "The term 'exercise' is an artifact of our industrialized, regimented, domesticated lives. If the brain is to take full advantage of what we now understand about the importance of movement, then you don't have to exercise; you've got to move. You've got to be nimble."[52]

Especially if you've been adversely affected by the bullying paradigm, choice is very important for you. Ratey and Manning are not trying to choose what movements you do, but like Merzenich, they hope for your brain's sake that you step off the treadmill and walk outdoors into nature. It's extremely good for the brain to be perpetually encountering the unpredictable world of nature: the uneven path, the cropping up of stones, the tree roots that surge out of the ground; the sounds, smells, wind, rain, sunlight, and creatures all contribute to vital brain systems that need to be constantly worked just like your muscles, tendons, and joints for optimal performance. As much as your brain needs your limbs moving at an aerobic pace, it also needs you to open up your senses to the natural world, navigate your way through it, and remember details when you return home. Likewise, it needs you to open your heart and connect with others. Your brain is profoundly and sensitively social. Take every chance you have to replace inactivity with movement. Why take an escalator when you can dash up the stairs? We think it's a bonus for modern citizens to be "supported in their life of ease by technological advances," but as Merzenich cautions, these devices are "designed to help you sleepwalk through life." And the more you sleepwalk, the more your brain declines as it's hungry for learning, for novelty, and for challenge.[53] Sleepwalking results when we insulate "ourselves from the real world."[54]

"Nimble" is a term that applies both to physical and cognitive movements. A person can have agile fingers and an agile mind. A nimble person can be quick in movement and quick in intellect. If we want to attain this kind of agility, it won't happen with exercise alone or mindfulness alone. As Merzenich advises, if we truly want to exit the bullying paradigm and fulfill our remarkable potential, we have to exercise the brain and invest in specific exercises that relate more directly to what's wrong.[55] In this way, brain training is more like physiotherapy that is required to make us healthy and whole *before* we launch into our fitness regime. Exercise unto itself is "not enough to normalize the brain of a kid who has been bullied or abused."[56] It's not their fault that they are unmotivated, despondent, passive, or cowering; that they have low self-esteem or lack belief in themselves. These are the barriers to exercise, *not* laziness. So maybe it's time that we tackled these barriers with an evidence-based, brain-training intervention that can heal traumatized brains and get them healthy, fit, and ready to enter the wild world of aerobic fitness.

When you exercise in nature, you are working with your body and brain as evolution designed them and that keeps them both nimble. "Simply by engaging the real, rocky, rolling world and its variety of stresses and varied challenges, you have engaged the full range of muscular and neurological activity that evolved in places like this."[57] The fulfillment you experience when you meet a physical and mental challenge translates into joy or bliss, which releases dopamine, nature's antidepressant, into your brain. "Evolution has made provisions for our happiness, but to take advantage of them you've got to move."[58] What I think was most healing for me post-abuse was that the relationship that first and foremost mattered was with myself. I didn't need anyone to exercise with me or tell me how to exercise. Instead of looking out into the world for a guide, I turned inward and listened to "the still, small voice inside. The voice of knowing." For me, the voice of knowing emerges from the harmonious dialogue among my Mind-Brain-Body. And that voice of knowing is like listening to many years of evolution that abruptly made the random teachers who abused me seem very minor in comparison. I wondered why I would give so much time, thought, and anguish to these interrupters, when instead I could be focused on and listening to my ancestors. When I'm moving, I tune into my ancestral self, sculpted to survive throughout time.

CHAPTER 10

# Listen to Your Mind-Brain-Body

## *Step 10: Hear Your Whole Voice*

In December 2015, I was approached by a business advisor and equity investor who had a dream to found a start-up that assisted college graduates in transitioning into the workplace more quickly and efficiently. He had been watching his friends' children and his own sons graduate only to find themselves either underemployed or unable to find even a starting position. There appeared to be a disconnect between professors and employers, the programs in colleges and the needs of the workplace, even when grads were coming out of business schools. The issue had reached crisis proportions in Canada so that 50 percent of college grads were spending two to five years in underemployment, and some did not find jobs at all. Surely we could do better for our young people. We had a series of discussions around having me research, develop curriculum, and then pilot a program the following year to see if there was enough interest. This was several months after Ellen had reported to me that she was being sexually harassed. I was deeply disillusioned with teaching in private schools and jumped at the chance to do something different.

I loved the idea of doing a deep dive into researching a field I was essentially ignorant about and then developing new curriculum to see if I could help. I knew it would be a very steep learning curve to enter into the intersectional world of career and business, but my passion for research and for learning made me hope that I would have something to contribute. My training in grad school had been to work with university students, and I had lectured at University of Toronto for years, so I felt truly excited about this opportunity. I greatly respected the business advisor and the small board he put together; we all wanted to contribute to the start-up, and I knew they would be insightful mentors. So I threw caution to the wind and took a year's leave of absence from teaching. We launched "C2Careers" the following autumn.

My visionary employer generously funded all kinds of courses for me, and I learned a great deal in a short period of time; however, there was one course that was life-changing. The course was offered by Lee-Anne Gray, a psychologist, author, and educational innovator in California. I had crossed paths with Gray two years earlier when *Teaching Bullies* was published as she had included it in an article she wrote for the *Huffington Post* on educational trauma.[1] After we discussed the issue of educators who use bullying as a way to "teach," Gray wrote a subsequent, more in-depth treatment of *Teaching Bullies*.[2] The subject was something that, as a psychologist, as a founder of a school, and as a mother of three, was close to her heart. It was also relevant to her brilliant mind, and in 2019, her latest book, *Educational Trauma*, was published. What made me turn to her in 2016 when I was trying to learn ways to assist college grads to find a foothold in the workforce was the motto for the school she had founded: "Empathic Education for a Compassionate Nation." Practically every book, blog, and article I read insisted that "empathy" was the crucial ingredient for success in career launching, in the workplace, and for leaders. Daniel Pink even sees it as a key ingredient for those who want to triumph in the future workplace. Pink connects empathy with storytelling, healing, and creating a holistic mind.[3] I contacted Lee-Anne to see if she might be able to teach me more about empathy, and that's how I ended up being trained by her in "empathic listening."

Session after session, Lee-Anne taught me the skill of empathic listening. She would set the intention of compassionate interaction. For empathy to work in a pro-social way, you need to move beyond "affect-matching."[4] You can walk in someone's shoes for a mile and then focus on how *your* feet hurt. You can watch people suffering on TV and quickly change the channel after an empathic reaction. If we want to harness empathy as a way to express more understanding and compassion, then we need to not just feel someone else's pain but consciously, caringly respond. We need to take time to hear the other person and remain focused on them and their needs, not our own. Once Lee-Anne provided the compassionate foundation, she could then proceed with the "empathic education." After compassionate intention was expressed, then we would decide who would speak first. In the first few sessions, I would speak first and learn from Lee-Anne how to listen empathically. I would tell her about something for a minute or two and then stop. She would not interject or respond in any way, not even nod her head or smile. Her only focus was on listening. Her job was to repeat back to me, with as much accuracy as possible, what I had said. She could not add anything to what I had said or subtract anything. The goal was hearing me and then repeating to me my own words. It was like an auditory mirror.

I might speak for five or seven minutes in this manner, stopping every minute or so to allow Lee-Anne to repeat what I had said. When I felt "heard," then I would announce exactly that "I feel heard." Pause for a moment and reflect on that sensation. Do you frequently feel "heard" in your life, or do you feel that no one is truly listening to you or people don't "get you" or when you want to speak, people interrupt you? Perhaps you've had the experience of someone speaking right over you or not responding at all to what you have said. The next reflection is on yourself. How well do you listen to others? Do you truly try to hear what they are saying and listen in silence until they are done? Do you interrupt or interject? Next time you are having a conversation, observe the pattern of your speaking and listening and notice how others speak and listen to you.

After Lee-Anne role-modeled for me her empathic listening so that I felt "heard," it was my turn and we reversed roles. Now my job was to repeat back to her everything she said. My first discovery was how hard it is to actually *listen* to someone else. I wanted to jump in with a confirmation, pose a question, insert a joke. And it's not permitted. In empathic listening, you cannot so much as make a sigh. You have to listen very closely to what the person says because you need to repeat back to them as accurately as possible, which takes great concentration. Much as you might want to, you must resist the impulse to add a word, thought, or alternative idea. Once you've repeated what they've said in the first minute or two, then they say more! And again, you need to repeat until the magical moment when they say, "I feel heard." And then finally, finally it is your turn to speak once again. On the one hand, empathic listening is different than a conversational back-and-forth; on the other hand, empathic listening means you might be having a more accurate exchange because you are actually, truly hearing one another and waiting your turn before sharing your own perspective.

Imagine using this technique when you are having a fight with your partner or having a disagreement at work. Psychiatrist Helen Reiss speaks of the importance of "hearing the whole person" when she works with couples in marital conflict.[5] Imagine using empathic listening when you are distraught about your teenager's decision or your elderly parent's refusal to move into assisted living. I promise you will discover whole layers, emotional depths, fears and griefs you could not have known existed. That's the beauty of truly listening and reflecting back to the person what they want to express rather than constantly asserting your own needs, thoughts, and feelings. Consciously practicing empathy, I realized, could operate as a powerful way to transform bullying into compassion. Empathy, as with all healthy qualities, starts with you and how you listen to yourself. If you want to feel whole and have your Mind-Brain-Body be aligned and integrated, empathic listening is an effective way. Psychiatrist Daniel Siegel writes about a patient who healed by developing body awareness, journal writing to explore his feelings, and using imagery or visualization to attend to nonverbal emotions. He notes that these "essential elements of empathy are all forms of integration."[6]

While bullying and abuse involve projecting onto a victim one's own feelings of self-loathing, shame, or fear, empathy involves listening to the way in which an individual experiences, feels, sees, thinks, and intends. While bullying involves projecting one's own history onto a victim to relieve one's own past trauma, empathy involves hearing an individual's history and imagining how it might shape their present. Empathy can be cognitive, whereby you decode another's experience using memory, logic, inference, and deduction, and empathy can be affective, whereby you perceive and interpret the world through the physical and emotional experiences of another.[7] Neuroimaging studies have confirmed that the parts of the brain engaged in thinking and feeling *connect* in empathic responses.[8] They have a holistic response. In contrast, bullying and abuse hinge on a lack empathy for another person in the sense that the one bullying or abusing fails to imagine, relate to, or respond to the suffering and humiliation being experienced by their victim. Phrases like "emotionally dysregulated," "unempathic," "callous unemotional" are used to describe children who bully, and these terms apply just as effectively to adults.[9] Bullying and empathy can be seen as polar opposites. Thanks to the technological advance in brain scans, we have an opportunity to see the way in which your brain grows, develops, functions, falters, and repairs. If you have been bullied or abused and find yourself trapped in self-harming or self-limiting behaviors, what you can do is unscramble the tangled wires in your brain.

Helen Reiss says that "empathy is best understood as a human capacity consisting of several different facets that work together to enable us to be moved by the plights and emotions of others."[10] In the 1990s, Italian neuroscientists discovered that if one primate or monkey observed another one eating something, then similar parts of its brain would activate. As noted, they named the brain cells that were firing up "mirror neurons" to convey the idea that one brain mirrors another. As neuroscientists continue to research the way in which one brain might mirror another, they have learned that for children empathy was a necessary skill for survival.[11] It's believed that to survive, infants and toddlers needed to learn what the adults in their world were thinking, feeling, and intending. Brain images have shown that we are indeed wired for empathy, or shared mind awareness, from the day we are born, and our empathic capacity requires specialized brain circuits that allow us to perceive, process, and respond to others.[12] Like all parts and structures in the brain, the empathy ones can be strengthened, weakened, or eliminated by the experiences you have and the practices to which you commit.

Decades before the discovery of "mirror neurons" in primates in the 1990s, psychoanalyst Heinz Kohut was studying what he called "mirroring transference." He examined the way in which children came to learn of their own uniqueness, their signature strengths, and their special qualities through the eyes of their caregivers when they reflected back to children these qualities. While Kohut did not know the brain had neural networks dedicated to this mirroring process, he observed this vital response for children when they had adults in their lives that met their eyes and met them with matching emotion, caring postures, and reflective facial expressions. Kohut referred to this as "psychological oxygen" for babies and children.[13] You can well imagine how in the bullying paradigm, some babies and children do not get this oxygen. They are neglected. No one mirrors to them their golden potential; instead, they may be exposed to noxious air that gives them a completely distorted idea of their potential, transforming the way they see themselves from gold to lead.

Neuroscientists revolutionized our understanding of empathy by not only studying one brain being hurt but instead studying two brains: one that was being hurt and one that was watching. Psychologist Tania Singer, the innovative researcher who set up this dual brain scanner experiment, is now a world leader in empathy and compassion. What was so remarkable in this first experiment was that the researchers discovered that the brains of the watchers essentially duplicated the experience of the people whose fingers were being stuck with needles and who were feeling actual pain. The watchers' brains responded as if they were feeling the pain themselves.

> Interestingly, the same neural networks light up during actual pain felt by those being pricked as during observation of the pinprick. When a part of your brain called the insular cortex fires, you experience the pain because there are neurons whose job it is to physiologically respond to pain. It turns out that a similar subset of neurons will also fire when you are a mere witness to an action that produces pain.[14]

The key difference is that when we feel pain, our brain zeroes in on this sensation as a survival strategy. In contrast, when we are witness to the pain of others, when we feel empathy with them—our brain makes it very clear to us what it would be like to have such pain—still, we focus on them, not ourselves. Our pain is not literal, signaling to us we are in danger and at risk of not surviving. Our "pain" is empathic and signals to us that we may need to help the one suffering. What Singer learned, and this research has been replicated many times, is that we have "mirror phenomena, or shared neural circuit mechanisms" in our brains that reflect, or communicate to us, what's happening in the brains of others.[15] This is why Reiss says that we have "shared mind intelligence" and "shared mind emotions or sensations or intentions."[16] We have shared brain circuits not only for pain or actions like eating but also for touch and feelings like disgust. Mirroring appears to involve brain regions that include the somatosensory cortex (receives information from the senses), the anterior cingulate cortex (implicated in empathy, emotion, impulse control, and decision-making), and the insula (which plays a role in pain perception, empathy, and social engagement, among other functions).[17] As Ratey and Manning explain:

> This consciousness of another's point of view is exactly what enables the more elegant and refined form of lying so valuable to all humans: storytelling. It allows abstraction and conceptualization, which in turn allows language. It allows a concept of the future, which in turn opens the door to planning and scheming and is why planning is related to empathy.[18]

Empathy is innate, whereas bullying and abuse are learned behaviors (mostly, except for people who are quite possibly born sociopaths or psychopaths). Merzenich weighs in with his expertise and wisdom at this juncture. He nuances the simplistic statement "empathy is innate" and explains that "we are born with the power to generate deep attachment and feeling for one another and are designed to reliably develop it." However, the key to remember throughout this book is that empathy evolves in our brain "through plastic change," and "life can be so challenging to frustrate its development."[19] He is quick to point out that if I say "empathy is innate," I may well place the "burden of shame on someone who never had the advantage of developing strong attachments to others." The word "yes" resonates through my whole being when he reminds me of this fundamental concept. I grew up with an old-fashioned saying from my Swedish grandfather and Scottish grandmother, and it stayed with me: "But for the grace of God go I." In other words, if someone has done a disastrous thing, you recognize that you too might have done such a thing and it's only good fortune, a great upbringing, a divine intervention that has saved you.

This kind of empathy and compassion dominates Merzenich's approach. If I am empathic, it is because I grew up in an empathic world. If I struggle with empathy, it is because I am likely traumatized from experiences in my formative years. As Merzenich shares, "I personally think that most bullies were neurologically wounded at a young age, and their bullying was a natural and predictable outcome of their brain-damaged circumstances."[20] Absolutely, critically important, and if this in fact is the case, think about how insane it is that instead of offering a diagnosis, intervention, or plan for a cure, we cover up abusive conduct and strive to make it disappear. It would be like realizing someone has cancer and covering it up. A doctor diagnosing diabetes or arthritis but then covering it up. Does this help the patient in the long run? What if the doctor diagnosed an infectious disease and then covered it up? Does that help the patient or protect those who are likely to become infected? Imagine how different our world would be if we identified and acted on diagnoses of "bullying and abuse" in all their forms and we intervened with a strategy for stopping the spread of the disease and striving to heal it. We read daily in the media what happens to perpetrators and victims when there's no diagnosis, no intervention, and no plan for healing, just cover-up until it is a full-on health crisis shattering the lives of far too many victims, the perpetrator(s), and those who were informed but did not act to protect. What happens in the brain when we stop covering up and we heal social-emotional illness and crisis?

Unsophisticated, aggressive behaviors displayed by alpha-male baboons were studied by neurologist Robert Sapolsky. These alpha males used a "more or less constant harassment of subordinates" to violently maintain their dominance. Sapolsky was interested in how this kind of treatment affected the subordinates, and it turned out they had very high levels of cortisol. More disturbing was the discovery that those in the British public service had comparably high levels.[21] If you've read the harrowing work of Alex Renton, on the way in which rampant abuse hammers the empathy out of British children in elite schools, creating an abuse cycle where victims become perpetrators, it's perhaps not surprising to find that the "ruling class," in positions of dominance over public servants, cause their cortisol levels to rise.[22]

A lethal disease struck the baboon colony that Sapolsky was studying, and it attacked the alpha males and eliminated most, if not all, of them. The subordinates did not re-create the harassing, aggressive society that had dominated them. Instead, they became more cooperative and caring. Their cortisol levels dropped.[23] Some individuals appear to be born with heightened empathy levels, while others have more muted empathy levels. Merzenich adds to this study an incredibly important insight. Doesn't this study suggest that alpha males look out for or have empathy for one another?[24] In the bullying paradigm, this is so true. It's hard to walk away from this framework because we are empathic with it. We walk a mile in the shoes of the perpetrators far more often than their victims. Because so many of us are flawed, are raised in the bullying paradigm, are aware of our own missteps where we have inadvertently or purposefully harmed others, we find that our understanding and protectiveness swings over to the abusive perpetrators, not the victims. The victims come to represent to our brains a rebel group that might one day point out our failings or our abusive moments. This fear stops us then from halting the abuse. It's a sort of empathy gone haywire. However, once the bullying is gone, once the alpha males are gone, "the bullied have empathy for one another."[25] In summary, empathy can make us align with perpetrators or victims. As an innate quality, it all depends on how it develops through our experiences as expressed in our brains' plastic change.

For some highly sensitive people, feeling empathy is natural and automatic. These empaths sometimes need to turn down the dial on their emotional empathy to become objective enough to do their jobs. Others are not highly empathic or sensitive to others, and they may need to develop their empathy skills. What's exciting is that neuroplasticity—our response to our environment but also our ability to purposely shape our brain—means that we can dial up or down our naturally occurring empathy.

Do not mix up empathy with being "nice" or "kind." It's more complex. Psychologist Matthew Lieberman's work on our social brain depicts the innate impulses we have to connect as three stages that essentially apply to a more nuanced understanding of empathy. His research documents that our social bonding responds to Connection, Mindreading, and Harmonizing. Connection begins with birth and is most intense for babies and children but stays with us all our lives. It hinges on our capacity, with other mammals, to "feel social pains and pleasures," thereby "forever linking our well-being to our social connectedness." The next level is Mindreading, which is our ability to understand the thoughts, intentions, and actions of those around us. This ability we have further bonds us as a group that can share ideas and think strategically. The third aspect comes into play in adolescence. Lieberman calls it Harmonizing because, as his research shows, teens develop a sense of self within the world of their peers. Their brains become highly attuned to how their peers see them and interact with them, and this creates greater "social cohesion."[26]

It's important not to mix up empathy with sympathy. Research has shown that sympathy can cause disconnection. When an individual feels sorry for someone and says, "That must be terrible," she is not walking in their shoes; if anything, she's "keeping a safe distance." Brené Brown distinguishes between the "powerful 'me too' of empathy" and the "'not me,' but I feel for you, of sympathy," adding that while empathy has healing power, sympathy is more likely to trigger shame.[27] Shaming, a veritable cornerstone of the bullying and abuse paradigm, makes us believe we are unworthy, rejected, and a nonperson. This belief then leads to the breaking apart of our Mind-Brain-Body. It shatters us and can lead to a lack of self-empathy and a lack of empathy for others. Daniel Siegel writes about a patient who essentially "took refuge in her cortex, to cut herself off from the ongoing pain of criticism, isolation and unfairness."[28] It's an apt description of the ways we can fragment ourselves. The patient was using the rational part of her brain, her prefrontal cortex, to suppress and avoid feelings from the limbic or emotional region of her brain. Keeping neuroplasticity in mind, if our brains can be "fragmented" in the way that they work, then we can change that by healing our scars and restoring our holistic health.

Economist and author Jeremy Rifkin describes in powerful terms how shaming works to break social bonds and unravel community. When we do something harmful and become aware that we have hurt someone, then the natural response is guilt that triggers "empathic distress and the desire to reach out and make amends."[29] In this way, empathy serves to create and maintain connection. In contrast, shame collapses and severs connection. Rifkin explains that shame "denigrates a person's being, making them feel worthless and inhuman. To be shamed is to be rejected. Shame is a way of isolating a person from the collective we. He or she becomes an outsider and a nonperson."[30] With Rifkin's description ringing in our minds, isn't it unbelievable that children are so often shamed for their lack of knowledge, their inexperience, their mistakes? What would our world be like if we responded to children's mistakes with empathy and compassion?

I want you to close your eyes and just do a quick check on yourself. Is it possible that within you there is a Mind-Bully who tells you you're "worthless" or "inhuman"? Is there a Mind-Bully who rejects you and makes you feel like an "outsider" or a "nonperson"? These are the kinds of shaming words and behaviors that rule the bullying paradigm. It's very important that you do not succumb to the Mind-Bully and shame your brain or body, as a repeat of shaming you may have endured in the past or present. The bullying paradigm deeply believes that shaming is the way to motivate or discipline, but the truth is that shaming your brain or body for errors, failures, weakness, or falling prey to harmful habits does *not* motivate you or "teach you a lesson." Shame does not make you feel like you belong and that you can take your rightful place in the collective we.

Instead, shame, whether being projected on you or as an emotion that you have internalized, only serves to entrench destructive neural networks that over time become a cycle of striving, achieving, faltering, slipping, crashing back into harmful behaviors, feeling ashamed, self-medicating in unhealthy ways, then starting all over again. This shame cycle simply reinforces your belief that you can never stop destructive behaviors or habits. The shame cycle makes you believe you will never get healthy or free from these limiting beliefs and behaviors. For people to commit sexist, racist, ageist, homophobic, anti-Semitic, misogynistic, or any other abusive acts that harm others, they first must turn off their natural empathy and believe that the one they attack, hurt, or demean is not worthy of respect, empathy, compassion, and love. So much suffering in our world occurs because we have been raised in a framework that condones bullying and abuse from childhood on, while vociferously denying it each step of the way.

Imagine how truly horrendous it is if indeed you might have this kind of cruel, taunting, harmful voice speaking within so that you don't even realize it as separate, false, and untrustworthy. It took me a long, long time, many hours with psychologists and psychiatrists, to truly hear my internalized Mind-Bully as false, destructive, and misleading. Having been brutally shamed as a teenage girl, for not succumbing to the three sexually abusive teachers in the Quest Program, I had my work cut out for me to fix the damage done. But I did fix it. And if you too have been bullied or abused, you can stop taking it out on yourself; you can become truly clear that you are worthy; you are a person *worthy* of empathy, respect, compassion, and love. The deadly force that shame has, especially when it is operating unnoticed within us, is that it can unwire our natural empathy for ourselves and others. Rifkin describes the way in which shame dismantles our natural empathy.

> Shame has the effect of turning off the innate empathic impulse. If one feels like a nonbeing, socially ostracized and without self-worth, he is unable to draw upon his empathic reserves to feel for another's plight. Unable to emotionally connect with others, he either shrinks into withdrawal or acts out his sense of abandonment by exercising rage at others.[31]

Well, that pretty much summarizes the bullying paradigm and explains how it operates as a cycle. The person who has an internalized Mind-Bully who relentlessly makes them feel worthless and strips them of empathy either harms herself with destructive habits or lashes out at others (or does both). Imagine all this happening within your skull and that much of this cyclic, destructive behavior is normalized in our world. The internalized Mind-Bully—which is all about shame, blame, and ostracize—is what turns off your brain's innate empathy.

The amazing thing is you were born with a brain wired for empathy, so if you lack it now in your life—finding yourself beset by feelings of worthlessness or feeling judgmental and angry, putting others down, maybe even raging a bit—then you might wonder what has happened to your brain. Remember when John Ratey talked about the impact of chronic stress on your brain and he said the hippocampus looked shrunken and shriveled? Well, it may manifest in your life as shrinking into "withdrawal." Rifkin expands from shaming an individual to shaming as a cultural paradigm or framework, like the one so many of us have been raised in and struggle with.

> Shaming cultures, throughout history, have been the most aggressive and violent because they lock up the empathic impulse, and with it the ability to experience another's plight and respond with acts of compassion. When a child grows up in a shaming culture believing that he must conform to an ideal of perfection or purity or suffer the wrath of the community, he is likely to judge everyone else by the same rigid, uncompromising standards.[32]

It's ominous to hear a line such as "believing he must conform to an ideal of perfection or purity" at a time when we are witnessing a surge of far-right ideology and fascism. To tap into a sense of belonging, they must reject the "collective we" and replace it with a "fragmented us vs. them." They blame, shame, and ostracize others as being "worthless" and "nonpersons." If we look at the culture these individuals are growing up in, do we find a shaming culture, a bullying culture whereby leaders in society publicly mistreat individuals and communities without any accountability or consequences? Are we shocked when these individuals resort to aggressive and violent thoughts and actions? As Mark Twain says, "History doesn't repeat itself, but it often rhymes," and considering the wars of the twentieth century, we would be fools to allow a shaming culture to dominate our homes, schools, sports, politics, and international relations. On an individual and a collective level, it is a matter of life and death to unlock the empathic impulse. We need to exit the bullying paradigm and enter an empathic paradigm whereby we "experience another's plight and respond with acts of compassion," whether that is self-compassion or compassion to a culture or nation in distress. It is time to unleash empathy so that we can return to our golden health and potential as individuals and as societies. Lieberman's research shows that our need for connection and community neurologically overrides even our need for food and shelter.[33]

Sadly, for many of us, we were raised, trained, and harmed in the bullying paradigm that makes us feel trapped in a shaming culture. Our natural, inborn empathic impulse is locked up behind bars begging to get out. That empathic response is what makes you cry when you see someone suffering; it's what makes you hear another's mind, heart, and soul so that you want to uphold their civil rights; it's what makes you stand up and protest when someone is being treated cruelly and with injustice. That empathy gets locked up in shaming cultures. With every single one of us who converts from shame to empathy, we slowly but surely create a new culture, one that is informed by neuroscience. And what does neuroscience tell us? Our brains are wired for empathy. Bullying and abuse are *unnatural* constructs that are imposed on the human brain. Bullying and abuse are *learned* behaviors. Neuroscience reminds us that we came into this world wired for empathy, and if it's been stripped from us, we need to rise up and take it back.

While the bullying paradigm under which we have lived for decades is heaving its last enraged breath, we are lurching in fits and starts into what Jeremy Rifkin terms the "Empathic Civilization." Surprisingly for an economist, Rifkin is a wonderful storyteller, and he shares the following tale in his book *The Empathic Civilization.*

> Several years ago, zoologists noticed a bizarre change in behavior among adolescent elephants in an animal park in South Africa. The young elephants began to taunt rhinos and other animals and even began to kill them, something never seen before. Scientists were puzzled by the strange behavior and unable to find a satisfying explanation. Then one of the zoologists recollected that years earlier they had culled out the older male elephants in order to ease overcrowding.

They reasoned that there might be a correlation, but they were not sure what it could be. Nonetheless, they airlifted two older male adults back into the park and within just a few weeks the teenagers stopped exhibiting what amounted to antisocial behavior and began to fall in line with the behavior of the older male elephants. What the zoologists observed is that young elephants learn from their elders, just like human children, and when the role models are absent they fail to learn what appropriate social behavior should be.[34]

If societies' leaders—whether at work, in school, in sports, or in politics—are bullying, then children will bully. If societies' leaders are empathic, then children will learn to behave like their elders. Research into empathy demonstrates there is hope for changing the culture of any individual or organization that prioritizes relationships, as demonstrated in the culture of elephants.

Tens of thousands of students have gone through the "Roots of Empathy" program that was begun by Mary Gordon in Toronto in 1996, and even children from abusive backgrounds responded very positively to empathy training. The bullying paradigm is quick to say that children do not have time to learn empathy because they are too busy with academics, which are much more important. However, what teachers found was that the "development of empathic skills leads to greater academic success in the classroom." As Mary Gordon says, "Love grows brains."[35] While that may sound emotional, Gordon's assertion is confirmed on brain scans. As Reiss explains: "Empathic capacity requires specialized brain circuits that allow us to perceive, process, and respond to others," and most importantly, "empathy *can* be taught."[36] While it is common knowledge that brains change shape and capacity with academic, musical, or sport training, it is less well-known that we can wire our brains to be more empathic and less reactive by how we conduct ourselves and what we focus our attention on.[37]

And empathy's effect on children, as well as adults, enhances learning on a deep level: "A troubled child is less likely to be an attentive and engaged learner than a happy child."[38] The less troubled you are, the more likely it is that you will be attentive, engaged, and a lifelong learner. You will be happier. Like mindful individuals, those who have empathy are not reactive. They create space between stimulus and response, as they factor in and strive to experience the feelings, thoughts, and intentions of others. In fact, research has shown just how "powerful the wiring in our brains is for pro-social, or helping behavior."[39]

Once again side-stepping "Descartes' Error," which tries to separate our thinking selves from our embodied, emotional selves, those who are trained in mindfulness and empathic thinking and listening are essentially being trained in critical thinking: "The ability to entertain conflicting feelings and thoughts, be comfortable with ambiguity, approach problems from a number of perspectives, and listen to another's point of view are essential emotional building blocks to engage in critical thinking."[40] The empathic mindset creates the collective we and depends on a foundation of many voices; the bullied or shamed mindset fragments the collective we and strives to silence facts, truths, and voices that it has relegated to the "out-group." The exciting and hopeful breakthrough with neuroscience is to discover that if you, or others you know, use bullying and abusive practices, you can *unlearn* them and, with repeat practice, can activate the specialized brain circuits correlated with empathy. Likewise, if you are on the receiving end of callous, unempathic conduct, then you need first and foremost to protect yourself, but second, hope that the individual can exit the bullying paradigm and enter into the empathic civilization.

The next time you're speaking with someone or a group, make a concerted effort to not only look into peoples' eyes but also note their eye colour.[41] Eyes are potent conveyers of peoples' feelings, and you will more quickly tap into the emotional atmosphere you're in by taking time to look into peoples' eyes. Educational experts and authors Desautels and McKnight draw on neuroanatomy and focus on the way in which trauma causes children to shut down and act out.[42] Children who grow up without support and intervention may well become adults who shut down and act out. While the cycle seems impossible to stop, the fact is it's never too late to reprogram the neuroanatomy, to rewire neural networks that are just not serving you anymore. Experts understand that your brain is a social organism, and if it's going to develop healthy relationships and resiliency, we need to get better informed about the neurobiological principles of social interaction and interpersonal relationships.[43] That's where empathy comes rushing in.

John Medina talks about empathy in illuminating ways. He refers to it as "Theory of Mind": the ability to understand the interior motivations of another and make predictions based on that knowledge.[44] This ability is what Lieberman refers to as "mindreading," and he sees it as one of the three core components of our social brains. Theory of Mind is a fundamental component of empathy because it allows for "perspective taking," which requires focused attention, imagination, and curiosity about all others, not just those with whom we share experiences or backgrounds.[45] Perspective taking is a conscious act that arises out of our natural empathic response. Neuroscientific research has shown that when people adopt the perspective of another, "the brain areas that are active in the first person become activated in the observer."[46] This is a process where we strive to understand a situation from another person's psychological, social, physical, and spiritual point of view. When the person is similar to us and thus easy to relate to, our perspective taking is more easily activated than when we do not feel we have much in common with someone—or have even gone as far as to put them into an "out-group." What's worrisome is that scientific research has shown that "there is an inverse relationship between power and empathy. This distance often insulates the rich and powerful from the suffering of the everyday individual."[47] If an individual is powerful and prone to put others into "out-groups," then empathy can be easily derailed. This is why removing power imbalance is so crucial. The greatest power imbalance lies between adults and children, and yet in the bullying paradigm, we do not put in nearly enough checks and balances to ensure that empathy has not been erased due to an adult's excessive power over kids. I can't help but wonder if the reason adults informed about abuse far too often protect the perpetrator and not the victims is because children have fallen into the "out-group" in terms of their empathy. In other words, the powerful adult relates to the other powerful adult who children report as abusive. The adult then focuses their empathy on the other adult. Merzenich pointed out this phenomenon with the bullying alpha-male baboons.

While shaming blocks an individual, a group, or our own brain and body from belonging to the "collective we," empathy is about integration and inclusion. "The more we relate directly to others and their lived experience, the more we can empathize," but we must tread carefully while relating because "the people we relate to the most are those we consider our 'in-group.'"[48] Thus, an excellent empathic exercise is to make a concerted effort to extend kindness and curiosity to individuals who are not culturally approved by your background, who are not acceptable to your in-group, who don't share your experiences. It's important to become aware that on an unconscious level, "we humans still act tribally." Scientists document that in-group bias "is so ingrained and often so subliminal that most of us struggle to be objective."[49] Do you notice how that sounds an awful lot like the internalized Mind-Bully? Take a moment to do a quick check: are there parts of yourself that you have relegated to the out-group, parts of yourself about which you feel shame? If so, it may well be that you've internalized the aggressor and have developed a lack of empathy for yourself. The Mind-Bully is sometimes so ingrained and operates so subliminally that we don't even know it's there. A healthy, useful practice is to invest time and effort in opening up, reflecting on, and most importantly relating to the out-group parts of your Mind-Brain-Body, while you extend this same curiosity and kindness to others you might assume are different from you. If you are an adult receiving reports of bullying and abuse by children, you must ensure that you offer them empathic understanding equal to, if not more than, what you offer to the accused adults.

You might feel as if strengthening empathy will compromise your competitive edge. Let me assure you, empathy is in fact proven to be a vital tool for achievement and triumph. It serves you well to reflect on, imagine, and focus on what your colleagues, allies, and competitors are feeling, and thus what they're planning and intending. And when the going gets tough, when the team is under siege, when the scoreboard is down, empathy works as a powerful motivator and forceful mechanism for pulling everyone together for a common cause. It supplies a bonding experience when you all recognize one another's stress and pain and go the next mile to alleviate it. Why? Because you're all pumped full of empathy. Empathic individuals are team players, and they want more than anything to make one another feel powerful, strong, and triumphant. Empathy is the glue that makes teams connect like a bonded family that nothing can pull asunder. Self-empathy can help you work through your mistakes and blocks more quickly and help you to refocus on the challenge at hand, thereby getting better results.

Shaming your own brain or body only causes you to dwell on your mistake—and feel fear and anxiety that you will make another mistake—rather than merely see an error as an opportunity to learn. If you put down or even berate your brain, you are causing corrosive stress hormones to pump into said brain, which hinders your ability to focus and problem solve. Ask yourself if you want to focus or be stressed and scattered. If the answer is *yes, you want to focus*, then bring on the empathy training. Blaming a colleague, teammate, family member, or friend when an error occurs or something goes wrong only serves to divide and fragment, thereby making everyone less connected. The task that lies before us is to unlock our natural empathy to unwire abusive and bullying neural networks.

Learning to activate your empathy can put you on a path of open-hearted courage where you learn to perceive what is healthy and whole, as well as what's fragmented and unhealthy. Instead of looking away or suppressing what's in the out-group, you discover that you can hold disparate emotions or judgments with courage, kindness, and curiosity and thereby halt the cycle of trauma.[50] Perspective taking through an empathic lens can also be applied to bullying and abuse itself: an individual can have many wonderful qualities (what's healthy and whole about them), but it does not mean that they are exempt from doing harm (the fragmented, unhealthy parts). Our brains, sequential processors that they are, struggle to hold these contradictory states, but informed by research, we know that individuals who bully and abuse almost always have many wonderful qualities. Oftentimes, they're intelligent, popular, even charismatic. They deserve respect and kindness if they are reported on as abusive, and they need to be given the opportunity to rehabilitate, but only if they are first held publicly accountable. Furthermore, they must not be in positions that allow them to repeat the abuse. After rehabilitation, it is crucial that they are monitored. I do not mean to suggest this is an easy process, and am well aware it depends on the severity of the abuse, how malleable their brain is, what stage they are at in life, how long the abuse has been repeated, and so forth. The point is the outdated blame-shame-ostracize model does not work. The paying of money to victims either to keep them quiet or to somehow compensate for the harm done does not work. We need to be more empathic. One of the best ways to dismantle the bullying and abuse framework and replace it with a new neuroparadigm is by rehabilitating those who bully and abuse. The sooner they are stopped, the easier the rewiring of their neural networks, the less victims are harmed, and the quicker the abuse cycle is halted.

Neuroscientists explain that working memory is crucial for empathy because for you to stop yourself from doing something harmful or destructive, you must be able to hold in your mind an alternative goal, someone else's point of view, or the future consequences of your present action.[51] When you practice empathy, you can think of it as exercising your working memory. If you do not train, practice, and work at empathy in your brain, you lose those neural paths. Just like if you don't exercise and remain passive, you lose muscle mass. You can strengthen your muscles by training them; likewise, you can strengthen desirable components of your brain function by training them.[52] Maps or networks in your brain wither with lack of use or grow and flourish by being regularly exercised.

That said, making the shift from bullying (or hearing the Mind-Bully) to empathy (or hearing the Empathic-Coach) is as difficult as committing to daily meditation practice or getting off the couch and committing to an exercise regime. The practice is not complex, nor does it require advanced instruction. It does not have to be expensive or even cost anything. Like meditation and exercise, it requires time, commitment, and practice. It's up to you, but I want to acknowledge how hard it is. If you've been bullied or abused, and you are tired of feeling ashamed, what looms before you is a mountain. Those who haven't been bullied or abused have a nice level playing field. But if you've internalized the harmful acts or words done or said to you, then that nice straightforward field becomes a vast mountain that takes a huge amount of energy, willpower, self-belief, courage, and outright fury to climb. Why fury? Because you know you are worthy, and somehow, someone, somewhere stripped that away from you. You need to fight to restore it.

"C2Careers" failed like so many start-ups do, but instead of demanding perfection and then beating myself up for failing to reach this ideal, I chose empathic listening as my big takeaway from the experience. What I learned from the business advisor and the board he put together was invaluable. The dream behind C2C was fueled by the strong impulse in the older generation to support, coach, and mentor the younger generation. These men were all devoted fathers, and they responded to the crisis of underemployment and unemployment with an empathic impulse to help. It is this deep-seated, evolutionary empathy that we are going to tap into in this step to still the chaos and silence the noise that fragments the Mind-Brain-Body after it has suffered bullying and abuse.

# Step 10: Hear Your Whole Voice

Remember Paul Madaule, who we discussed in chapter 4? He was the child in France who had learning disabilities, or cognitive deficits, and was unwittingly or quite purposely bullied by the adults in his world who thought he wasn't trying and was simply "lazy." He dedicated his life to saving kids who like him had learning exceptionalities and needed innovative approaches to allow them to flourish in their learning. In the many schools he has founded, Madaule has developed ways to heal children who suffer from various brain injuries and challenges, and one of the ways is by using recordings of the mother's voice. This concept intrigued me because I thought it could be applied to the "noise" that bombards the bullied and abused brain. If words could harm and scar the brain, couldn't they also be used to soothe and support the brain?

Biomechanics expert Joaquin Farias has come to realize in his work that dystonias—which are involuntary movements like spasms caused by overuse—can also result from "psychological trauma." He explains that "mental and physical traumas can trigger 'brain shock.'"[53] That is the clearest description of what I felt as a girl when my teacher publicly humiliated me with the label "frigid" when I recoiled at the idea of slow-dancing with him. That caused a shock to my brain, but I didn't know at the time, or for decades afterward. I couldn't heal the blow to my brain because I didn't know it had even happened. In Merzenich's research, he and his team have learned that healing focal dystonias from repeated movements, like those made by musicians or factory workers, occurs not by physiotherapy with the hand but by harnessing neuroplasticity and changing the brain. Moreover, this approach—"driving a brain, through training, in corrective directions, back toward normalcy—could also be applied to mental health conditions such as trauma and even schizophrenia."[54]

I decided that first, I needed to visualize what had happened to my brain, and then I could put together a recording for my brain. My goal was to silence what had been said and done to me and then replace that "noise" with my own empathic voice. It's an auditory version of Cognitive Behavioral Therapy whereby you consciously identify and replace "negative self-talk" with caring and compassionate words.

As a sixteen-year-old at my first Quest dance, my teacher, Dean Hull, asked me to slow-dance with him. I was horrified. When I said no, he labeled me "frigid" in front of my peers. Initially, I resisted, but then when he humiliated me in front of the other teachers and my fellow students and friends, I succumbed to his sickening embrace as a way to discount what he called me. I felt myself fragment. One part of me was slow-dancing with my teacher, while another part was utterly frozen in shock and fear, and yet another part was suffused with self-loathing and shame. From that day forward, I became withdrawn, physically awkward, and ashamed. I felt that there was something wrong with me and hoped no one would discover it. This was merely one moment within years of being expected to hug these teachers, share details about boys with them, give them massages or receive massages from them, and so on. This was merely one moment in a daily onslaught of sexual comments, public shaming, and giving privileges to those who succumbed to their sexual abuse and humiliation to those who resisted. I couldn't even hug my own mother without feelings of anxiety and distress by the time they were done with me. The public humiliation when I was an adolescent was scarring and so traumatizing I blanked it right out for years. I just assumed that I was an untouchable, physically awkward, chilly sort of individual.

How could the one word "frigid" have so much power over me?

Neuroscientist Norman Doidge quotes French otolaryngologist Alfred Tomatis, who explains that language possesses a physical dimension: "By causing vibrations in the surrounding air, language becomes a sort of invisible arm by which we 'touch' the person listening to us in every sense of the term." Even when an infant or child does not understand the meaning of language, his brain is still able to pick up the "emotional charge" of the messaging.[55] This is why during childhood we are so susceptible to words spoken to us, labels affixed to us, and also the tone of voice used by the adults in positions of power and authority over us—namely, parents, other family members, teachers, religious leaders, doctors, and coaches. Tone of voice conveys almost 40 percent of the nonverbal, emotional content of what a person communicates.[56] If adults in positions of trust abused that trust, and actually used words and tone of voice *not* to care for you, guide you, and support you in the fulfilling of your potential, but instead put you down, berated, demeaned, threatened, groomed, or shamed you, then you need to work hard to silence this harmful legacy.

If you were bullied or abused, like I was, then you might be paralyzed by the words spoken, and it's important to realize from a brain perspective that you might still be trapped in a state of fear. One way to recover from this paralyzed state or "broken-record" way of being is to activate your empathy. The empathy needs to be directed at your brain. I returned to the lessons taught to me by Lee-Anne Gray. I began conducting sessions of empathic listening, not with other people but within my own self. I set myself the task of listening empathically to my Mind and then to my Brain and then to my Body. It's a mindfulness exercise really, but instead of striving to stay clear of thoughts and feelings, I let them speak and I simply listened without judgment and with kindness and curiosity. It took a certain amount of courage, I discovered.

As Helen Reiss explains, from a brain perspective, or neurobiologically speaking, empathy requires "tamping down your own amygdala-driven threat sensors while listening to the other person."[57] The key word is "listening." Empathic listening makes it possible to connect. Let's rewrite Reiss's statement to apply it to a conversation between three characters, namely your Mind, your Brain, and your Body. In this case, empathy requires tamping down your own amygdala-driven threat sensors (in your Brain) while listening to the other part of you (your Mind and Body). In this scenario, you are *listening* to your Brain, which is having panicky reactions and is no longer being present. The Brain's alarm center, the amygdala, is firing on all cylinders, believing you—just like you were hurt or shamed as a kid—are about to be hurt or shamed again. You need to use your Mind to talk to your Brain and Body with great kindness; tell your Brain and Body it's okay, you're grown-up now, powerful, safe, and no longer dependent on adults who may have breached their positions of trust. Use a gentle tone of voice. Ignite your empathy and shower it on your poor Brain and Body that may well have been badly mistreated in the past.

While Madaule uses recordings of the mother's voice or of classical music, you could use a recording of your own voice as a way to reprogram how you speak to yourself. You could use the voice of a friend or family member, the voice of someone you love. For me though, the one voice I had not listened to nearly enough in my life was what George Mumford calls "the still, small voice within. The voice of knowing."

Tune in, not to just the words but to the tone of voice you are using. "We humans are exquisitely sensitive to variations in tone of voice and its prosody. When you say about someone, 'He's really good at that . . .' how you say it telegraphs the meaning. Did you convey admiration, sarcasm, contempt, surprise, fear, or disgust?"[58] If you want to speak empathically to your Mind-Brain-Body, then its crucial to focus on tone of voice, which is "more important than the actual words we say and can determine whether there is an empathic communication." When I interviewed Reiss, I was struck by her beautiful voice that resonates with empathy. When you speak with her, she consciously meets your eyes, gazes at you, sees you, and then radiates through her facial expression and body language that she hears you, the whole you.

# Conclusion
## THE NEW *NEUROPARADIGM*

Writing a book is paradoxically an experience of isolation and community. While it requires hours, months, and even years alone as the author grapples with others' research, ideas, insights, and conclusions, these very figures enter into the writer's own mind. They can be mentors, influential and supportive of one's own experiences, thoughts, and feelings; or they can be critics, divergent and provocative, making one question one's own experiences, thoughts, and feelings. The textual community that I have gathered in this book has provided extensive evidence to show us that we are so indoctrinated by the bullying paradigm that we "bully" our own brains. We ignore them. We ostracize them as if they have nothing to add to the conversation. For far too many of us, the brain is banished. The second we realize this odd truth about our present-day society, and of course it is a generalization, we can change.

The overarching goal of *The Bullied Brain: Heal Your Scars and Restore Your Health*, regardless of what you take away from this book, is that at least you have entered into a dialogue with your brain front and center. Not limiting yourself to the concept of mind/body and instead replacing that outdated idea with a holistic, aligned Mind-Brain-Body is unto itself a breakthrough. Once seeing yourself in this way, it becomes possible to examine how these three inextricably related aspects of yourself interact. Are they working at cross-purposes? Does one take precedence over the others? Is one aspect hurt or damaged? Can the other aspects of selfhood be harnessed to set in motion healing?

Many of us have been raised and trained, arguably indoctrinated, in an outdated bullying paradigm that no longer serves us. Research is clear that our brains have just as much, if not more, capacity to get healthier, more flexible, stronger, and quicker if we exercise them like we exercise our bodies. Our harmed parts can be integrated and brought back into a healthier, happier, calmer whole when we practice mindfulness and empathic listening. Wherever you are on the bullying spectrum—one who abuses, one who bullies, one who was or is a victim, one who witnesses bullying, one who looks away, one who has suppressed the trauma, one who speaks up, one who advocates on behalf of those harmed, one who takes the hard path of whistleblowing and strives to shine a spotlight on the damage being done and the system that supports it—you can change if you want. It's your choice. It's your brain, and you have the power to shape it and sculpt it. Never forget: your brain has the built-in ability to change. There is no single strategy that brings about healing. It must be a multipronged approach. Brain training designed by neuroscientists targets weak parts of our brains and strengthens them. Meditation calms us. Exercise reduces harmful stress. Empathic listening creates community and self-awareness. However, none of these strategies alone is a cure-all. The key is, if we want to restore all of the operational powers of our brains that may have been compromised by chronic stress resulting from bullying and abuse, then we need multiple strategies.

It appears that our bullying epidemic is not cured by blaming, shaming, and ostracizing. It is time for a revolution in how we think about and how we educate ourselves and children about bullying. It needs to be seen as a medical, not a moral, problem, especially in childhood. It needs to be instantly identified. The perpetrator needs to be held accountable. Healing needs to begin. The victim's brain needs to be assessed and treated, just like her body would be if it was physically sick or injured. We need to understand that even if they look whole from the outside, within their skull, serious damage may continue to plague them and even threaten their lives. The inspiring, evidence-based fact is that the vast majority of what is damaged in the brain by bullying and abuse can be repaired. It is time to integrate evidence-based brain training into our daily lives to keep our brains healthy and to optimize their performance. For those who are struggling with a specific brain distortion, neuroscientists' brain training is available now to treat it. We have an evidence-based treatment for many brain ailments, certainly for the ones caused by bullying and abuse.

Reaching the conclusion of *The Bullied Brain: Heal Your Scars and Restore Your Health* brings with it a particular bittersweet ending as it draws to a close my dialogue with Dr. Michael Merzenich. While I don't want the conversation to end, I also feel an intense responsibility to share his incredible knowledge of the brain's neuroplasticity as quickly as possible. His final thoughts on the book are insightful as always.

> We live in a world plagued by bullies. They come in all sizes and ages, and can plague us from the day we arrive on the planet to the day we die. They damage our brains (if we let them). They HAVE damaged brains. We now have the tools in hand to overcome this neurological trauma and distortion in both the bullied, and the bully. We can train our brains to restore their organic health and operational

functionality to a high level. We can calm our demons and bring our thinking brain back into control, by employing meditation and other highly useful emotion and attention directed changes. We can engage our physical bodies in healing forms of exercise. We can restore the empathetic powers of our bullied brains, and of the bully's brain.

One of the most amazing aspects of Merzenich's brain is that he is adept in the realm of deep science, where few of us can even go, and yet he has an ability to synthesize and articulate ideas in such a way that we can all learn from him. He sees in *The Bullied Brain* the key message, which is that we can be part of "a new revolution where bullying transforms into HEALING, for bullied and bully, alike."

In a poignant final plea, Merzenich addresses those who bully and abuse: "Just stop it. With neurological strengthening, you can stop. With a recognition of the destructiveness of your bullying and abusive behaviors on the innocent children and adults around you, just stop. Heal thyself."

And then he addresses me, and all of us who believe that we can have a more neuro-informed, kinder, more insightful, more compassionate, more high-functioning, more balanced, more holistic, healthier sense of self and sense of our world: "Help them heal themselves. Imagine a world where we all cared for one another. Truly. Let's begin to make one."[1] I don't know about you, but when one of the world's leading neuroscientists issues a call to action about creating a world based on mutual care, I am inspired and I am heeding the call.

Neuropsychologist Rick Hanson shares a childhood memory of standing across the street from his house, alone in the dark, when he was six years old.[2] At this same age, my brother came home and said, "I think my brain is crippled." Children speak in ways that can break your heart.

In Hanson's childhood neighborhood, it had recently rained, and he felt "sad about the unhappiness that night" in his home. If you close your eyes for a moment, maybe you can feel what it would be like to be outside in the dark. No one seems to know you're gone, and whatever it is that's going on at home, you don't want to be there. It's such a lonely and lost moment.

After teaching his readers amazing lessons constructed on the intersection between neuroscience and Buddhism, Hanson returns again to his childhood sadness, and we learn that it wasn't just that one night. Instead, it appears the unhappiness in his home and his sadness colored his whole childhood. Hanson says it created "a hole" in his heart.

It's painful to read this because he has just explained that "self networks" in the brain get activated when you're "threatened or unsupported." He has just advised that you make sure your fundamental needs are met. Stressing that we "all need to feel cherished," you can't help but wonder if he's learned this lesson the hard way.

Hanson explains: "Empathy, praise, and love from others—especially in childhood—are internalized in neural networks that support feelings of confidence and worth." His mention of childhood signals to me once again that while he is an expert in the brain, he still can feel very sharply the unhappiness of that six-year-old from so long ago. He adds, almost as if addressing that boy standing out in the dark, looking in sadness at his unhappy home, that when *you* do not get empathy, praise, and love in childhood then "you're likely to end up with a hole in your heart." On the next page, Hanson shares that the hole in his own young heart was "as big as the excavation for a skyscraper."[3]

When he was a boy who felt sad about the unhappiness in his home, on that dark night after it rained and he was alone, Hanson looked away from his present situation and focused on "the distant hills" where "tiny lights twinkled." And here is the moment where you can see his brain at work: "Then it came to me very powerfully: it was up to me, and no one else, to find my way over time toward those faraway lights and the possibility of happiness they represented."[4] It's almost as if Hanson sees the eighty-six billion glittering neurons full of light and electricity in his own brain. Because it is a choice and we can rewire our neural networks, he takes the adversity of his childhood and transforms it into a joyful life. Striving throughout his life to move away from unhappiness and toward happiness, he works hard through scientific observation, positive psychology, and meditation practice to heal the hole in his heart. The very fact that he teaches meditation and writes as a brain expert about the ways in which we can wire happiness into our brains means that it's something we all can do.

Hanson had a massive hole in his heart, opened up by adversity in childhood, but by working with his Mind-Brain-Body, he healed and restored his health. He teaches:

> No matter how big your own hole is, each day hands you at least a few bricks for it. Pay attention to good things about yourself and the caring and acknowledgement of others—and then take them in. No single brick will eliminate that hole. But if you keep at it, day by day, brick by brick, you'll truly fill it up.[5]

The bricks represent moments of happiness, moments when you give your best effort and don't give up on yourself, moments when you listen to scientific research, moments when you commit to training your body and your brain, moments when you make choices based on evidence-based practices, moments when you believe in yourself, moments when you connect with others, moments when you're treated with empathy and you strive to offer it to all, moments when you're honored with praise and cherished for your uniqueness, moments when you're heard, moments when you tap into flow, moments when you move aerobically in nature, moments when you are mindful, moments when you let go of any labels falsely applied to you and instead listen to the still, quiet voice within—the voice of knowing. Those are the moments of the new *neuroparadigm*, and while they are meant to heal the hole in your heart, they're also designed to open up the remarkable glittering landscape of your brain.

# Notes

## Introduction

1. Nicolas Burra, Dirk Kerzel, David Munoz Tord, Didier Grandjean, and Leonardo Cerevolo, "Early Spatial Attention Deployment Toward and Away from Aggressive Voice," *Social Cognitive and Affective Neuroscience* 14, no. 1 (January 2019): 73–80, https://doi.org/10.1093/scan/nsy100.

2. Rick Hanson, *Hardwiring Happiness: The New Brain Science of Contentment, Calm, and Confidence* (New York: Penguin, 2013), 23.

3. John Medina, *Brain Rules: 12 Principles for Surviving and Thriving at Work, Home, and School* (Seattle: Pear Press, 2008), 186.

4. David Walsh, *Why Do They Act That Way? A Survival Guide to the Adolescent Brain for You and Your Teen* (New York: Simon & Schuster, 2004), 95.

5. Daniel Christoffel, Sam Golden, and Scott Russo, "Structural and Synaptic Plasticity in Stress-Related Disorders," *Nature Reviews Neuroscience* 22, no. 5 (2011): 535–49, 10.1515/RNS.2011.044.

6. Tracy Vaillancourt, Eric Duku, Suzanna Becker, Louise Schmidt, Jeffrey Nicol, Cameron Muir, and Harriet MacMillan, "Peer Victimization, Depressive Symptoms, and High Salivary Cortisol Predict Poor Memory in Children," *Brain and Cognition* 77 (2011): 191–99, https://mimm.mcmaster.ca/publications/pdfs/s2.0-S0278262611001217-main.pdf.

7. Medina, *Brain Rules*, 178.

8. Gabor Maté, *In the Realm of Hungry Ghosts: Close Encounters with Addiction*, revised edition (Toronto: Penguin, 2018), 34.

9. Stan Rodski, *Neuroscience of Mindfulness: The Astonishing Science Behind How Everyday Hobbies Help You Relax, Work More Efficiently and Lead a Healthier Life* (New York: HarperCollins, 2019), 17.

10. Medina, *Brain Rules*, 178.

11. Jacqui Plumb, Kelly Bush, and Sonia Kersevich, "Trauma-Sensitive Schools: An Evidence-Based Approach," *School Social Work Journal* (2016), https://www.semantic

185

scholar.org/paper/Trauma-Sensitive-Schools%3A-An-Evidence-Based-Plumb-Bush/39c
27626fdef81b93b57eccfc41309772dbc6f78.

12. Sarah-Jayne Blakemore, *Inventing Ourselves: The Secret Life of the Teenage Brain* (New York: Hachette, 2018), 38–39.

13. Daniel Amen, *Change Your Brain, Change Your Life: The Breakthrough Program for Conquering Anxiety, Depression, Obsessiveness, Lack of Focus, Anger, and Memory Problems*, revised edition (New York: Penguin, 2015), 12.

14. Amen, *Change Your Brain*, 22.

15. James Clear, *Atomic Habits: An Easy and Proven Way to Build Good Habits and Break Bad Ones* (New York: Penguin, 2018), 19.

16. Lori Ward and Jamie Strashin, "Sex Offences against Minors: Investigation Reveals More Than 200 Canadian Coaches Convicted in the Last Twenty Years," *CBC*, February 10, 2019, https://www.cbc.ca/sports/amateur-sports-coaches-sexual-offences -minors-1.5006609.

17. Alexander Wolff, "Why Does Women's Basketball Have So Many Coaching Abuse Problems?" *Sports Illustrated*, October 1, 2015, https://www.si.com/college/2015 /10/01/abusive-coaches-womens-basketball-illinois-matt-bollant.

18. Steve Reilly, "Teachers Who Sexually Abuse Students Still Find Classroom Jobs," *USA Today*, December 22, 2016, https://www.usatoday.com/story/news/2016/12/22 /teachers-who-sexually-abuse-students-still-find-classroom-jobs/95346790/.

19. Bonnie Stiernberg, "USA Gymnastics Culture of Abuse Runs Far Deeper Than Larry Nassar," *Inside Hook*, July 20, 2020, https://www.insidehook.com/article/sports /usa-gymnasticss-history-of-abuse.

20. Andrew Sapakoff, "College of Charleston Report Hammers 'Jekyll and Hyde' Verbal Abuse by Coach Doug Wojcik," *Post and Courier*, July 2, 2014, https:// www.postandcourier.com/sports/college-of-charleston-report-hammers-jekyll-and-hyde -verbal-abuse-by-coach-doug-wojcik/article_6be0a402-b99b-5db3-8d0f-4e2b1ea12e4e .html.

21. Jeannie Blaylock, "Public Can Check Boy Scout 'Perversion Files' for Accused Molesters," *First Coast News*, November 12, 2019, https://www.firstcoastnews.com /article/news/investigations/boy-scouts-sexual-abuse-investigation/77-6b587579-410a -4f37-a9a0-d7f3cb0000b6.

22. ESPN, multiple contributors, "Inside a Toxic Culture at Maryland Football," *ESPN*, August 10, 2018, https://www.espn.com/college-football/story/_/id/24342005 /maryland-terrapins-football-culture-toxic-coach-dj-durkin.

23. Laura Clementson and Gillian Findlay, "'It's Overwhelming': Survivors Create Public List of Catholic Clerics Accused of Sexual Abuse," *CBC*, December 5, 2019, https://www.cbc.ca/news/canada/catholic-sexual-abuse-london-diocese-1.5384217.

24. BBC, "Canada: 751 Unmarked Graves Found at Residential School," *BBC*, June 24, 2021, https://www.bbc.com/news/world-us-canada-57592243.

25. Alan McEvoy and Molly Smith, "Statistically Speaking: Teacher Bullying Is a Real Phenomenon, but It's Been Hard to Quantify—Until Now," *Teaching Tolerance Magazine* 58 (Spring 2018), https://www.tolerance.org/magazine/spring-2018 /statistically-speaking.

26. Alan McEvoy, "Abuse of Power: Most Bullying Prevention Is Aimed at Students. What Happens When Adults Are the Aggressors?" *Teaching Tolerance Magazine* 48 (Fall 2014), https://www.tolerance.org/magazine/fall-2014/abuse-of-power.

27. Paul Pelletier, *The Workplace Bullying Handbook: How to Identify, Prevent and Stop a Workplace Bully* (Vancouver: Diversity Publishing, 2018), 110.

28. Pelletier, *Workplace Bullying*, 109.

29. Brené Brown, *Rising Strong: How the Ability to Reset Transforms the Way We Live, Love, Parent, and Lead* (New York: Random House, 2015), xxi.

30. Brown, *Rising Strong*, xviii.

31. McEvoy and Smith, "Statistically Speaking."

32. Helen Reiss, *The Empathy Effect: Seven Neuroscience-Based Keys for Transforming the Way We Live, Love, Work, and Connect across Differences* (Boulder, CO: Sounds True, 2018), 64.

33. Jennifer Fraser, *Teaching Bullies: Zero Tolerance on the Court or in the Classroom* (Vancouver: Motion Press), 2015.

34. Ciceri, in conversation.

35. Merzenich, "Childhood" (unpublished manuscript), 84.

36. Merzenich, "Childhood" (unpublished manuscript), 84.

37. Merzenich, "Childhood" (unpublished manuscript), 86.

38. Merzenich, e-mail correspondence.

39. Michael Merzenich, *Soft-Wired: How the New Science of Brain Plasticity Can Change Your Life* (San Francisco: Parnassus Publishing, 2013), 153–66.

40. David Cooperson, *The Holocaust Lessons on Compassionate Parenting and Child Corporal Punishment* (self-published via CreateSpace, 2014), 18.

41. Lee-Anne Gray, *Educational Trauma: Examples from Testing to the School-to-Prison Pipeline* (London: Palgrave Macmillan, 2019), 177.

42. Alex Renton, *Stiff Upper Lip: Secrets, Crimes and the Schooling of a Ruling Class* (London: Weidenfeld & Nicholson, 2017).

43. Paul Axelrod, "Banning the Strap: The End of Corporal Punishment in Canadian Schools," *EdCan*, January 6, 2011, https://www.edcan.ca/articles/banning-the-strap-the-end-of-corporal-punishment-in-canadian-schools/.

44. Elizabeth Gershoff, Andrew Grogan-Kaylor, Jennifer Lansford, Lei Chang, Arnaldo Zelli, Kirby Deater-Deckard, and Kenneth Dodge, "Parent Discipline Practices in an International Sample: Associations with Child Behaviors and Moderation by Perceived Normativeness," *Child Development* 81, no. 2 (March 2010): 487–502, https://www.ncbi.nlm.nih.gov/pmc/articles/PMC2888480/pdf/nihms-198378.pdf; Gray, *Educational Trauma*, 177.

45. Jeremy Rifkin, *Empathic Civilization: The Race to Global Consciousness in a World in Crisis* (New York: Penguin, 2010), 117.

46. Gray, *Educational Trauma*.

47. Laurence Steinberg, *Age of Opportunity: Lessons from the New Science of Adolescence* (New York: Houghton Mifflin Harcourt, 2014), 11.

48. Molly Castelloe, "How Spanking Harms the Brain: Why Spanking Should Be Outlawed," *Psychology Today*, February 12, 2012, https://www.psychologytoday.com/ca/blog/the-me-in-we/201202/how-spanking-harms-the-brain.

49. Renton, *Stiff Upper Lip*.

50. Frank Larøi, Neil Thomas, André Aleman, Charles Fernyhough, Sam Wilkinson, Felicity Deamer, and Simon McCarthy-Jones, "The Ice in Voices: Understanding Negative Content in Auditory Verbal Hallucinations," *Clinical Psychology Review* 67 (February 2019): 1–10, https://doi.org/10.1016/j.cpr.2018.11.001.

51. Bill Hathaway, "Past Abuse Leads to Loss of Gray Matter in Brains of Adolescents," *Yale News*, December 5, 2011, https://news.yale.edu/2011/12/05/past-abuse-leads-loss-gray-matter-brains-adolescents-0.

52. Matthew Lieberman, *Social: Why Our Brains Are Wired to Connect* (New York: Random House, 2013), 68–69.

53. Bessel van der Kolk, *The Body Keeps the Score: Brain, Mind, and Body in the Healing of Trauma* (New York: Penguin, 2015), 168.

54. Gray, *Educational Trauma*, 29.

55. Joe Dispenza, *Breaking the Habit of Being Yourself: How to Lose Your Mind and Create a New One* (Carlsbad, CA: Hay House, 2012); Daniel Reisel, "The Neuroscience of Restorative Justice," *TED Talk*, February 2013, https://www.ted.com/talks/dan_reisel_the_neuroscience_of_restorative_justice?language=en.

56. Sofia Bahena, North Cooc, Rachel Currie-Rubin, Paul Kuttner, and Monica Ng, eds., *Disrupting the School-to-Prison Pipeline* (Boston: Harvard Educational Review, 2012).

57. Amen, *Change Your Brain*, 15.

58. Reisel, "Neuroscience of Restorative Justice."

59. Amen, *Change Your Brain*, 31.

60. Merzenich, *Soft-Wired*, 32.

61. Norman Doidge, *The Brain's Way of Healing: Remarkable Discoveries and Recoveries from the Frontiers of Neuroplasticity* (New York: Penguin, 2016), xix.

62. Doidge, *Brain's Way of Healing*, xx.

63. Doidge, *Brain's Way of Healing*, xix.

64. Rachel Nuwer, "Coaching Can Make or Break an Olympic Athlete: Competitors at the Most Elite Level Need More than Technical Support," *Scientific American*, August 5, 2015, https://www.scientificamerican.com/article/coaching-can-make-or-break-an-olympic-athlete/.

65. Sarah-Jayne Blakemore and Uta Frith, *The Learning Brain: Lessons for Education* (Malden, MA: Wiley-Blackwell, 2005), 3.

66. Robert Cribb, "Teachers' Bullying Scarred Us Say Student Athletes," *Toronto Star*, March 14, 2015, https://www.thestar.com/news/canada/2015/03/14/teachers-bullying-scarred-us-say-student-athletes.html; CTV W5, "Personal Foul: Sports Dreams Shattered by Aggressive Coaches," *CTV W5*, March 14, 2015, https://www.ctvnews.ca/video?clipId=569994&playlistId=1.2279107&binId=1.811589&playlistPageNum=1&binPageNum=1.

# Chapter 1

1. Renata Caine and Geoffrey Caine, "Understanding a Brain-Based Approach to Learning and Teaching," *ASCD*, 1990, http://www.ascd.org/ASCD/pdf/journals /ed_lead/el_199010_caine.pdf, 67.

2. Emily Anthes, "Inside the Bullied Brain: The Alarming Neuroscience of Taunting," *Boston Globe*, November 28, 2010, http://archive.boston.com/bostonglobe/ideas /articles/2010/11/28/inside_the_bullied_brain/, 3.

3. Matthew Lieberman, *Social: Why Our Brains Are Wired to Connect* (New York: Random House, 2013), 45–70.

4. James Clear, *Atomic Habits: An Easy and Proven Way to Build Good Habits and Break Bad Ones* (New York: Penguin, 2018), 1–7.

5. Anthes, "Bullied Brain," 3.

6. Michael Merzenich, *Soft-Wired: How the New Science of Brain Plasticity Can Change Your Life* (San Francisco: Parnassus Publishing, 2013), 22.

7. Merzenich, *Soft-Wired*, 22.

8. Merzenich, e-mail correspondence.

9. Merzenich, *Soft-Wired*, 22.

10. Anthes, "Bullied Brain," 1.

11. Alex Renton, *Stiff Upper Lip: Secrets, Crimes and the Schooling of a Ruling Class* (London: Weidenfeld & Nicholson, 2017).

12. Anthes, "Bullied Brain," 1.

13. Sarah-Jayne Blakemore, *Inventing Ourselves: The Secret Life of the Teenage Brain* (New York: Hachette, 2018); Frances Jensen and Amy Ellis Nutt, *The Teenage Brain: A Neuroscientist's Survival Guide to Raising Adolescents and Young Adults* (Toronto: HarperCollins, 2015); David Walsh, *Why Do They Act That Way? A Survival Guide to the Adolescent Brain for You and Your Teen* (New York: Simon & Schuster, 2004); Daniel Siegel, *Brainstorm: The Power and Purpose of the Teenage Brain* (New York: Penguin, 2013); Laurence Steinberg, *Age of Opportunity: Lessons from the New Science of Adolescence* (New York: Houghton Mifflin Harcourt, 2014).

14. Anthes, "Bullied Brain," 2.

15. Kimberly Archie, Solomon Brannan, Tiffani Bright, Jo Cornell, Debbie Pyka, Mary Seau, Cyndy Feasel, Marcia Jenkins, Leanne Pozzobon, and Darren Hamblin, *Brain Damaged: Two-Minute Warning for Parents* (Westlake Village, CA: USA Sport Safety Publishing, 2019).

16. Merzenich, e-mail correspondence.

17. Stanley Greenspan, with Beryl Benderly, *The Growth of the Mind: And the Endangered Origins of Intelligence* (New York: Perseus Books, 1997), 252–53.

18. Merzenich, e-mail correspondence.

19. Anthes, "Bullied Brain," 1.

20. Anthes, "Bullied Brain," 1.

21. Thomas Kuhn, *The Structure of Scientific Revolutions*, third edition (Chicago: University of Chicago Press, 1996).

22. Merzenich, *Soft-Wired*, 2.

23. Norman Doidge, *The Brain's Way of Healing: Remarkable Discoveries and Recoveries from the Frontiers of Neuroplasticity* (New York: Penguin, 2016), 353.

24. Doidge, *Brain's Way of Healing*, 355.

25. Anthes, "Bullied Brain," 1.

26. Rick Hanson and Richard Mendius, *Buddha's Brain: The Practical Neuroscience of Happiness, Love, and Wisdom* (Oakland, CA: New Harbinger, 2009), 42.

27. Helen Reiss, *The Empathy Effect: Seven Neuroscience-Based Keys for Transforming the Way We Live, Love, Work, and Connect across Differences* (Boulder, CO: Sounds True, 2018), 68.

28. Jensen and Nutt, *Teenage Brain*, 179.

29. Stan Rodski, *Neuroscience of Mindfulness: The Astonishing Science Behind How Everyday Hobbies Help You Relax, Work More Efficiently and Lead a Healthier Life* (New York: HarperCollins, 2019), 14–15.

30. Doidge, *Brain's Way of Healing*, 111.

31. John Ratey, *Spark: The Revolutionary New Science of Exercise and the Brain* (New York: Little, Brown and Company, 2008), 59.

32. Merzenich, *Soft-Wired*, 5.

33. Blakemore, *Inventing Ourselves*, 81.

34. Merzenich, e-mail correspondence.

35. Merzenich, *Soft-Wired*, 121.

36. Todd Sampson, "Redesign My Brain," *IMDb*, October 2013, https://www.imdb.com/title/tt3322570/episodes?year=2013&ref_=tt_eps_yr_2013.

37. John Arden, *Rewire Your Brain: Think Your Way to a Better Life* (Hoboken, NJ: John Wiley and Sons, 2010), 10.

# Chapter 2

1. Patricia Bauer, "Damien Chazelle: American Director and Screenwriter," *Encyclopedia Britannica*, last updated January 15, 2021, https://www.britannica.com/biography/Damien-Chazelle.

2. Damien Chazelle, "Before Writing and Directing 'Whiplash,' Damien Chazelle Lived It," *Los Angeles Times*, December 18, 2014, https://www.latimes.com/entertainment/envelope/la-et-mn-whiplash-writers-damien-chazelle-20141218-story.html.

3. A. A. Dowd, "*Whiplash* Maestro Damien Chazelle on Drumming, Directing, and J. K. Simmons," *AVClub*, October 15, 2014. https://film.avclub.com/whiplash-maestro-damien-chazelle-on-drumming-directing-1798273033.

4. Damien Chazelle, "Divide and Conquer: Damien Chazelle on Why You Should Make a Short First," *MovieMaker*, October 9, 2015, https://www.moviemaker.com/damien-chazelle-on-why-you-should-make-a-short-first/.

5. Stanton Pruitt, "Damien Chazelle's Films and the Consequences of Ambition," *Cultured Vultures*, October 7, 2019, https://culturedvultures.com/damien-chazelles-films-and-the-consequences-of-ambition/.

6. Oliver Gettel, "*Whiplash* Director Damien Chazelle on his Real-Life Inspiration," *Los Angeles Times*, November 11, 2014, https://www.latimes.com/entertainment/movies/moviesnow/la-et-mn-whiplash-damien-chazelle-real-life-inspiration-20141111-story.html.

7. Tasha Robinson, "Damien Chazelle on What Is and What Isn't Ambiguous in *Whiplash*," *The Dissolve*, October 15, 2014, https://thedissolve.com/features/emerging/787-damien-chazelle-on-what-is-and-isnt-ambiguous-abou/.

8. Chazelle, "Divide and Conquer."

9. Chazelle, "Divide and Conquer."

10. Charlie Schmidlin, "Interview: Director Damien Chazelle talks 'Whiplash,' Musical Editing and His New 'MGM-style' Musical 'La La Land,'" *IndieWire*, October 10, 2014, https://www.indiewire.com/2014/10/interview-director-damien-chazelle-talks-whiplash-musical-editing-his-mgm-style-musical-la-la-land-271422/.

11. Sarah-Jayne Blakemore, *Inventing Ourselves: The Secret Life of the Teenage Brain* (New York: Hachette, 2018), 43.

12. Don Kaye, "Interview: *Whiplash* Director Damien Chazelle," *Den of Geek*, October 9, 2014, https://www.denofgeek.com/movies/interview-whiplash-director-damien-chazelle/.

13. Eat Drink Films, "The Language of Drums: Director Damien Chazelle and Metallica's Lars Ulrich discuss *Whiplash*," *Eat Drink Films*, November 13, 2014, https://eatdrinkfilms.com/2014/11/13/the-language-of-drums-director-damien-chazelle-and-metallicas-lars-ulrich-discuss-whiplash/.

14. Dowd, "*Whiplash* Maestro."

15. Ashley Lee, "'Whiplash': J. K. Simmons, Damien Chazelle, on Whether Torment Leads to Talent," *Billboard*, September 27, 2014, https://www.billboard.com/articles/news/6266541/whiplash-jk-simmons-damien-chazelle-on-whether-torment-leads-to-talent.

16. Fred Topel, "Whiplash: Damien Chazelle on Sadistic Writing," *Mandatory*, October 6, 2014, https://www.mandatory.com/fun/770139-whiplash-damien-chazelle-sadistic-writing.

17. Chazelle, "Before Writing and Directing."

18. Conrad Quilty-Harper, "Damien Chazelle: My Next Film Will Have Less Cymbal Throwing," *GQ*, May 11, 2015, https://www.gq-magazine.co.uk/article/damien-chazelle-interview-whiplash-movie-jazz.

19. Blakemore, *Inventing Ourselves*, 87.

20. Bessel van der Kolk, *The Body Keeps the Score: Brain, Mind, and Body in the Healing of Trauma* (New York: Penguin, 2015), 102.

21. Robinson, "Damien Chazelle."

22. Robinson, "Damien Chazelle."

23. Ian Pace, "Music Teacher Sentenced to 11 Years in Prison as Abuse Film Whiplash Prepares for Oscars," *The Conversation*, February 20, 2015, https://theconversation.com/music-teacher-sentenced-to-11-years-in-prison-as-abuse-film-whiplash-prepares-for-oscars-37786.

24. J. Bryan Lowder, "Wailing Against the Pansies: Homophobia in *Whiplash*," *Slate Magazine*, October, 22, 2014, https://slate.com/human-interest/2014/10/why-does-whiplash-damien-chazelles-jazz-movie-contain-so-much-homophobia.html.

25. John Medina, *Brain Rules: 12 Principles for Surviving and Thriving at Work, Home, and School* (Seattle: Pear Press, 2008), 45–46.

26. John Ratey, *Spark: The Revolutionary New Science of Exercise and the Brain* (New York: Little, Brown and Company, 2008), 74.

27. David Sims, "The Uncomfortable Message in *Whiplash*'s Dazzling Finale." *The Atlantic*, October 22, 2014, https://www.theatlantic.com/entertainment/archive/2014/10/the-ethics-of-whiplash/381636/.

28. Andrew Sapakoff, "College of Charleston Report Hammers 'Jekyll and Hyde' Verbal Abuse by Coach Doug Wojcik," *Post and Courier*, July 2, 2014, https://www.postandcourier.com/sports/college-of-charleston-report-hammers-jekyll-and-hyde-verbal-abuse-by-coach-doug-wojcik/article_6be0a402-b99b-5db3-8d0f-4e2b1ea12e4e.html; Amber Jamieson, "Pace Football Coach Accused of Vicious Abuse by Players," *New York Post*, November 23, 2014, https://nypost.com/2014/11/23/pace-football-coach-abused-players-ex-team-members/; Heather Dinich, "Power, Control and Legacy: Bob Knight's Last Days at IU," *ESPN*, November 29, 2018, https://www.espn.com/mens-college-basketball/story/_/id/23017830/bob-knight-indiana-hoosiers-firing-lesson-college-coaches.

29. John Taylor, "Behind the Veil: Inside the Mind of Men Who Abuse," *Psychology Today*, February 5, 2013, https://www.psychologytoday.com/us/blog/the-reality-corner/201302/behind-the-veil-inside-the-mind-men-who-abuse.

30. A. O. Scott, "Drill Sergeant in the Music Room," *New York Times*, October 10, 2014, https://www.nytimes.com/2014/10/10/movies/in-whiplash-a-young-jazz-drummer-vs-his-teacher.html.

31. Merzenich, e-mail correspondence.

32. William Copeland, Dieter Wolke, Adrian Angold, and Jane Costello, "Adult Psychiatric Outcomes of Bullying and Being Bullied by Peers in Childhood and Adolescence," *JAMA Psychiatry* 70, no. 4 (2013): 419–26, doi:10.1001/jamapsychiatry.2013.504.

33. Helen Reiss, *The Empathy Effect: Seven Neuroscience-Based Keys for Transforming the Way We Live, Love, Work, and Connect across Differences* (Boulder, CO: Sounds True, 2018), 120.

34. Joseph Burgo, "All Bullies Are Narcissists: Stories of Bullying and Hazing in the News Break Down to Narcissism and Insecurity," *The Atlantic*, November 14, 2013, https://www.theatlantic.com/health/archive/2013/11/all-bullies-are-narcissists/281407/.

35. Stanley Greenspan, with Beryl Benderly, *The Growth of the Mind: And the Endangered Origins of Intelligence* (New York: Perseus Books, 1997), 52.

36. Frank George and Dan Short, "The Cognitive Neuroscience of Narcissism," *Journal of Brain Behavior and Cognitive Sciences* 1, no. 6 (2018).

37. George and Short, "Neuroscience of Narcissism."

38. George and Short, "Neuroscience of Narcissism."

39. George and Short, "Neuroscience of Narcissism."

40. William Verbeke, Vim Rietdijk, Wouter van den Berg, Roeland Dietvorst, Loek Worm, and Richard Bagozzi, "The Making of the Machiavellian Brain: A Structural MRI Analysis," *Journal of Neuroscience, Psychology and Economics* 4, no. 4 (2011), 10.1037/a0025802, 205–6.

41. Verbeke et al., "Machiavellian Brain," 212–13.

42. Paul Babiak and Robert Hare, *Snakes in Suits: When Psychopaths Go to Work* (New York: HarperCollins, 2007).

43. Paul Pelletier, *The Workplace Bullying Handbook: How to Identify, Prevent and Stop a Workplace Bully* (Vancouver: Diversity Publishing, 2018), 119.

44. Martin Teicher, "Wounds That Won't Heal: The Neurobiology of Child Abuse," *Cerebrum: The Dana Forum on Brain Science* 4, no. 2 (January 2000): 50–67, https://www.researchgate.net/publication/215768752_Wounds_that_time_won't_heal_The_neurobiology_of_child_abuse.

45. Teicher, "Wounds that Won't Heal."

46. Yvon Delville, Richard Melloni, and Craig Ferris, "Behavioral and Neurobiological Consequences of Social Subjugation During Puberty in Golden Hamsters," *The Journal of Neuroscience* 18, no. 7 (1998): 2667–72, https://doi.org/10.1523/JNEUROSCI.18-07-02667.1998.

47. Chazelle, "Before Writing and Directing."

48. Sims, "Uncomfortable Message."

49. Sean Fitz-Gerald, "Ask a Julliard Professor: How Real Is *Whiplash*?" *Vulture*, October 17, 2014, http://www.vulture.com/2014/10/ask-an-expert-juilliard-professor-whiplash.html.

50. Fitz-Gerald, "Ask a Julliard Professor."

51. Robinson, "Damien Chazelle."

52. Robinson, "Damien Chazelle."

53. Roger Rubin, Michael O'Keefe, Christian Red, and Nathaniel Vinton, "Mike Rice's Assistant Coach at Rutgers, Jimmy Martelli Resigns, Following Physical and Verbal Abuse Scandal," *New York Daily News*, April 5, 2013, http://www.nydailynews.com/sports/college/rutgers-assistant-baby-rice-cooked-article-1.1308334.

54. Reiss, *Empathy Effect*, 84.

55. Misia Gervis and Nicola Dunn, "The Emotional Abuse of Elite Child Athletes by Their Coaches," *Child Abuse Review* 13, no. 3 (June 24, 2004), https://onlinelibrary.wiley.com/doi/abs/10.1002/car.843; Carol Dweck, *Mindset: The New Psychology of Success* (New York: Ballantine, 2006); Pelletier, *Workplace Bullying*.

56. Movie Gal, "Interview with 'Whiplash' Film Maker and Oscar Nominee Damien Chazelle," *TheMovieGal.com*, February 8, 2015, https://www.themoviegal.com/single-post/2015/02/08/interview-with-whiplash-filmmaker-oscar-nominee-damien-chazelle.

57. Alex Renton, *Stiff Upper Lip: Secrets, Crimes and the Schooling of a Ruling Class* (London: Weidenfeld & Nicholson, 2017).

58. Erin Smith, Ali Diab, Bill Wilkerson, Walter Dawson, Kunmi Sobowale, Charles Reynolds, Michael Berk et al., "A Brain Capital Grand Strategy: Toward Economic Reimagination," *Molecular Psychiatry* 26 (October 2020): 3–22, https://www.nature.com/articles/s41380-020-00918-w.

59. Damien Chazelle, "Six Film-making Tips from Damien Chazelle: The 'La La Land' Director on How to Make it in La La Land," *Film School Rejects*, December 7, 2016, https://filmschoolrejects.com/6-filmmaking-tips-from-damien-chazelle-6f05f 190f427/.

# Chapter 3

1. Saul McLeod, "The Milgram Shock Experiment," *Simply Psychology*, updated 2017, https://www.simplypsychology.org/milgram.html. All references to Milgram's experiment that follow come from this article unless specified.

2. Cari Romm, "Rethinking One of Psychology's Most Infamous Experiments," *The Atlantic*, January 28, 2015, https://www.theatlantic.com/health/archive/2015/01 /rethinking-one-of-psychologys-most-infamous-experiments/384913/.

3. Gregorio Encina, "Milgram's Experiment on Obedience to Authority," *University of California*, November 15, 2004, https://nature.berkeley.edu/ucce50/ag-labor/7article /article35.htm.

4. Encina, "Milgram's Experiment."

5. Encina, "Milgram's Experiment."

6. Encina, "Milgram's Experiment."

7. Romm, "Rethinking."

8. Romm, "Rethinking."

9. McLeod, "Milgram Shock Experiment."

10. Norman Doidge, *The Brain That Changes Itself: Stories of Personal Triumph from the Frontiers of Brain Science* (New York: Penguin, 2007), 305.

11. Michael Merzenich, *Soft-Wired: How the New Science of Brain Plasticity Can Change Your Life* (San Francisco: Parnassus Publishing, 2013), 79.

12. Helen Reiss, *The Empathy Effect: Seven Neuroscience-Based Keys for Transforming the Way We Live, Love, Work, and Connect across Differences* (Boulder, CO: Sounds True, 2018), 29.

13. Reiss, *Empathy Effect*, 30.

14. Merzenich, e-mail correspondence.

15. Doidge, *Brain That Changes Itself*, xiv.

16. Gabor Maté, *In the Realm of Hungry Ghosts: Close Encounters with Addiction*, revised edition (Toronto: Penguin, 2018), 183.

17. David Eagleman, *Livewired: The Inside Story of the Ever-Changing Brain* (Toronto: Doubleday, 2020), 4.

18. Sarah-Jayne Blakemore, *Inventing Ourselves: The Secret Life of the Teenage Brain* (New York: Hachette, 2018), 61.

19. Eagleman, *Livewired*, 8.

20. Maté, *Hungry Ghosts*, 183.

21. Shawn Achor, *The Happiness Advantage: The Seven Principles of Positive Psychology That Fuel Success and Performance at Work* (New York: Random House, 2010), 167.

22. Achor, *Happiness Advantage*, 156.

23. Norman Doidge, *The Brain's Way of Healing: Remarkable Discoveries and Recoveries from the Frontiers of Neuroplasticity* (New York: Penguin, 2016), 11.

24. Merzenich, *Soft-Wired*, 58.

25. Angela Duckworth, *Grit: The Power of Passion and Perseverance* (New York: HarperCollins, 2016), 191–92.

26. Carol Dweck, *Mindset: The New Psychology of Success* (New York: Ballantine, 2006), 180.

27. Dweck, *Mindset*, 172.

28. Dweck, *Mindset*, 172.

29. Doidge, *Brain That Changes Itself*, 305.

30. Reiss, *Empathy Effect*, 117.

31. Merzenich, *Soft-Wired*, 63.

32. Merzenich, *Soft-Wired*, 62.

33. Dweck, *Mindset*, 184.

34. Mine Conkbayir, *Early Childhood and Neuroscience: Theory, Research and Implications for Practice* (London: Bloomsbury Academic, 2017).

35. Daniel Coyle, *The Talent Code: Greatness Isn't Born. It's Grown. Here's How* (New York: Random House, 2009), 162.

36. Merzenich, *Soft-Wired*, 63.

37. Coyle, *Talent Code*, 168.

38. Coyle, *Talent Code*, 171.

# Chapter 4

1. John Medina, *Brain Rules: 12 Principles for Surviving and Thriving at Work, Home, and School* (Seattle: Pear Press, 2008), 172.

2. Rick Hanson, *Hardwiring Happiness: The New Brain Science of Contentment, Calm, and Confidence* (New York: Penguin. 2013), 26.

3. Lee-Anne Gray, *Educational Trauma: Examples from Testing to the School-to-Prison Pipeline* (London: Palgrave Macmillan, 2019), 153.

4. Angela Duckworth, *Grit: The Power of Passion and Perseverance* (New York: HarperCollins, 2016), 190.

5. Sarah-Jayne Blakemore, *Inventing Ourselves: The Secret Life of the Teenage Brain* (New York: Hachette, 2018), 185.

6. Steve Silberman, *NeuroTribes: The Legacy of Autism and the Future of Neurodiversity* (New York: Penguin, 2015).

7. Alison Gopnik, *The Gardener and the Carpenter: What the New Science of Child Development Tells Us about the Relationship between Parents and Children* (New York: Farrar, Straus & Giroux, 2016).

8. Barbara Arrowsmith-Young, *The Woman Who Changed Her Brain: Unlocking the Extraordinary Potential of the Human Mind* (New York: Free Press, 2012).

9. Norman Doidge, *The Brain That Changes Itself: Stories of Personal Triumph from the Frontiers of Brain Science* (New York: Penguin, 2007), 41.

10. Doidge, *Brain That Changes Itself*, 39–40.

11. Michael Merzenich, *Soft-Wired: How the New Science of Brain Plasticity Can Change Your Life* (San Francisco: Parnassus Publishing, 2013), 80.

12. Merzenich, *Soft-Wired*, 79.

13. John Corcoran, *The Teacher Who Couldn't Read: One Man's Triumph over Illiteracy* (New York: Kaplan Publishing, 2008).

14. Merzenich, *Soft-Wired*, 29.

15. Doidge, *Brain That Changes Itself*, 41.

16. Norman Doidge, *The Brain's Way of Healing: Remarkable Discoveries and Recoveries from the Frontiers of Neuroplasticity* (New York: Penguin, 2016), 283.

17. Paul Pelletier, *The Workplace Bullying Handbook: How to Identify, Prevent and Stop a Workplace Bully* (Vancouver: Diversity Publishing, 2018), 94.

18. Medina, *Brain Rules*, 172.

19. Gabor Maté, *In the Realm of the Hungry Ghosts: Close Encounters with Addiction*, revised edition (Toronto: Penguin, 2018), 33.

20. David Eagleman, *Livewired: The Inside Story of the Ever-Changing Brain* (Toronto: Doubleday, 2020), 24.

21. Britt Andreatta, "Potential," *TEDx Talk*, July 22, 2014, https://www.youtube.com/watch?v=yXt_70Ak670&ab_channel=TEDxTalks.

22. Merzenich, *Soft-Wired*, 37.

23. Merzenich, *Soft-Wired*, 80.

24. Merzenich, *Soft-Wired*, 80.

25. Maté, *Hungry Ghosts*, 50.

26. Brené Brown, *Rising Strong: How the Ability to Reset Transforms the Way We Live, Love, Parent, and Lead* (New York: Random House, 2015), 46.

27. John Arden, *Rewire Your Brain: Think Your Way to a Better Life* (Hoboken, NJ: John Wiley & Sons, 2010), 10.

28. Arden, *Rewire Your Brain*, 9.

29. Arden, *Rewire Your Brain*, 9.

30. Arden, *Rewire Your Brain*, 9.

31. Hanson, *Hardwiring Happiness*, 111.

32. Arden, *Rewire Your Brain*, 9.

33. Maté, *Hungry Ghosts*, 8.

34. John Ratey, *Spark: The Revolutionary New Science of Exercise and the Brain* (New York: Little, Brown and Company, 2008), 40.

35. David Walsh, *Why Do They Act That Way? A Survival Guide to the Adolescent Brain for You and Your Teen* (New York: Simon & Schuster, 2004), 95.

36. Merzenich, e-mail correspondence.

# Chapter 5

1. Alex Renton, *Stiff Upper Lip: Secrets, Crimes, and the Schooling of a Ruling Class* (London: Weidenfeld & Nicholson, 2017).

2. I corrected minor grammatical errors in the Facebook post.

3. William Copeland, Dieter Wolke, Adrian Angold, and Jane Costello, "Adult Psychiatric Outcomes of Bullying and Being Bullied by Peers in Childhood and Adolescence," *JAMA Psychiatry* 70, no. 4 (2013): 419–26, doi:10.1001/jamapsychiatry.2013.504.

4. Emily McNally, Paz Luncsford, and Mary Armanios, "Long Telomeres and Cancer Risk: The Price of Cellular Immortality," *The Journal of Clinical Investigation* 129, no. 9 (2019): 3474–81, 10.1172/JCI120851.

5. Merzenich, e-mail correspondence.

6. Bessel van der Kolk, *The Body Keeps the Score: Brain, Mind, and Body in the Healing of Trauma* (New York: Penguin, 2015), 98.

7. Van der Kolk, *Body Keeps the Score*, 99.

8. Van der Kolk, *Body Keeps the Score*, 96–97.

9. Roland Summit, "The Child Sexual Abuse Accommodation Syndrome," *Child Abuse and Neglect* 7 (1983): 177–93, https://www.abusewatch.net/Child%20Sexual%20Abuse%20Accommodation%20Syndrome.pdf; Alice Miller, *For Your Own Good: Hidden Cruelty in Child-Rearing and the Roots of Violence* (New York: Farrar, Straus and Giroux, 1983).

10. Van der Kolk, *Body Keeps the Score*, 204.

11. Van der Kolk, *Body Keeps the Score*, 204.

12. Van der Kolk, *Body Keeps the Score*, 143.

13. Van der Kolk, *Body Keeps the Score*, 134.

14. Van der Kolk, *Body Keeps the Score*, 133.

15. Van der Kolk, *Body Keeps the Score*, 133.

16. Shawn Achor, *The Happiness Advantage: The Seven Principles of Positive Psychology That Fuel Success and Performance at Work* (New York: Random House, 2010), 94.

17. Achor, *Happiness Advantage*, 95.

18. Martin Teicher, "Impact of Childhood Maltreatment on Brain Development and the Critical Importance of Distinguishing between the Maltreated and Non-Maltreated Diagnostic Subtypes," *International Society for Neurofeedback and Research* (September 2017), https://drteicher.files.wordpress.com/2017/11/isnr_2017_keynote_teicher.pdf.

19. Teicher, "Childhood Maltreatment," 46.

20. Teicher, "Childhood Maltreatment," 46–47.

21. Martin Teicher, "Wounds That Won't Heal: The Neurobiology of Child Abuse," *Cerebrum: The Dana Forum on Brain Science* 4, no. 2 (January 2000): 50–67, https://www.researchgate.net/publication/215768752_Wounds_that_time_won't_heal_The_neurobiology_of_child_abuse; Andrew Burke and Klaus Miczek. "Stress in Adolescence and Drugs of Abuse in Rodent Models: Role of Dopamine, CRF, and HPA Axis," *Psychopharmacology* 231, no. 8 (2014): 1557–80, 10.1007/s00213-013-3369-1.

22. Lori Desautels and Michael McKnight, *Eyes Are Never Quiet: Listening Beneath the Behaviors of Our Most Troubled Students* (Deadwood, OR: Wyatt-MacKenzie Publishing, 2019).

23. Teicher, "Wounds That Won't Heal."

24. Jennifer Fraser, *Be a Good Soldier: Children's Grief in English Modernist Novels* (Toronto: University of Toronto Press, 2011), 25.

# Chapter 6

1. Jaak Panksepp, *Affective Neuroscience: The Foundations of Human and Animal Emotions* (Oxford: Oxford University Press, 1998), 57.

2. Robert Anda, Vincent Felitti, James Bremner, John Walker, Charles Whitfield, Bruce Perry, Shanta Dube, and Wayne Giles, "The Enduring Effects of Abuse and Related Adverse Experiences in Childhood," *European Archives of Psychiatry and Clinical Neuroscience* 256, no. 3 (April 2006): 174–86, https://www.researchgate.net/publication/275971785_The_Enduring_Effects_of_Abuse_and_Related_Adverse_Experiences_in_Childhood_A_Convergence_of_Evidence_from_Neurobiology_and_Epidemiology.

3. Alison Gopnik, *The Gardener and the Carpenter: What the New Science of Child Development Tells Us about the Relationship between Parents and Children* (New York: Farrar, Straus & Giroux, 2016), 20.

4. Merzenich, e-mail correspondence.

5. Anda et al., "Enduring Effects of Abuse."

6. Gabor Maté, *In the Realm of Hungry Ghosts: Close Encounters with Addiction*, revised edition (Toronto: Penguin, 2018), 181.

7. Maté, *Hungry Ghosts*, 181.

8. Merzenich, e-mail correspondence.

9. Merzenich, e-mail correspondence.

10. CDC, "Adverse Childhood Experiences (ACEs)," *Centers for Disease Control and Prevention*, https://www.cdc.gov/violenceprevention/aces/index.html.

11. Laura Starecheski, "Take the ACE Quiz and Learn What it Does—And Doesn't Mean," *NPR*, March 2, 2015, https://www.npr.org/sections/health-shots/2015/03/02/387007941/take-the-ace-quiz-and-learn-what-it-does-and-doesnt-mean, is the reference for all ACEs questions quoted in this section.

12. Merzenich, *Brain Dead* (unpublished manuscript), 23.

13. Antonio Damasio, *Descartes' Error: Emotion, Reason and the Human Brain* (New York: Penguin, 2005).

14. Vincent Felitti, "Reverse Alchemy in Childhood: Turning Gold into Lead," *Family Violence Prevention Fund* 8, no. 1 (Summer 2001): 1–4, http://akhouse.org/tarr/docs/HCR21_Position-Paper_Reverse-Alchemy-in-Childhood_V-Felitti.pdf. All following references to Felitti are to this article, unless specified otherwise.

15. Bessel van der Kolk, *The Body Keeps the Score: Brain, Mind, and Body in the Healing of Trauma* (New York: Penguin, 2015), 148.

16. Merzenich, in conversation.

17. Merzenich, *Brain Dead* (unpublished manuscript), 36.

18. Merzenich, *Brain Dead* (unpublished manuscript), 45.

19. Merzenich, in conversation.

20. Merzenich, *Brain Dead* (unpublished manuscript), 58.

21. Merzenich, in conversation.

22. Merzenich, e-mail correspondence.

23. Merzenich, *Brain Dead* (unpublished manuscript), 58.

24. Merzenich, e-mail correspondence.

25. Al Aynsley-Green, *The British Betrayal of Childhood: Challenging Uncomfortable Truths and Bringing about Change* (London: Routledge, 2019), 3.

26. Australian Government, "Rewire the Brain," *Try, Test, and Learn Initiative at the Ministry of Social Services*, https://www.dss.gov.au/rewire-the-brain.

27. Rick Hanson, *Hardwiring Happiness: The New Brain Science of Contentment, Calm, and Confidence* (New York: Penguin, 2013), xxvi.

28. Hanson, *Hardwiring Happiness*, xxvi.

# Chapter 7

1. Daniel Lang, "The Bank Drama: Four Hostages Were Taken During a Bank Robbery in Stockholm, Sweden in 1973. How Did They Come to Sympathize with their Captors?" *The New Yorker*, November 25, 1974, https://www.newyorker.com/magazine/1974/11/25/the-bank-drama. All following references to the Kreditbank robbery are to this article unless otherwise specified.

2. Charles Bachand and Nikki Djak, "Stockholm Syndrome in Athletics: A Paradox," *Children Australia* 43, no. 3 (June 2018): 1–6, https://doi.org/10.1017/cha.2018.31.

3. Alex Renton, *Stiff Upper Lip: Secrets, Crimes and the Schooling of a Ruling Class* (London: Weidenfeld & Nicholson, 2017).

4. Kathryn Westcott, "What Is Stockholm Syndrome?" *BBC News Magazine*, August 21, 2013, https://www.bbc.com/news/magazine-22447726.

5. All the following references on the Hearst kidnapping originate on the FBI website unless otherwise specified.

6. Shirley Jülich, "Stockholm Syndrome and Child Sexual Abuse," *Journal of Child Sexual Abuse* 14, no. 3 (2004): 107–29.

7. Jülich, "Stockholm Syndrome and Child Sexual Abuse."

8. Bessel van der Kolk, *The Body Keeps the Score: Brain, Mind, and Body in the Healing of Trauma* (New York: Penguin, 2015), 129–30.

9. Chris Cantor and John Price, "Traumatic Entrapment, Appeasement and Complex Post-Traumatic Stress Disorder: Evolutionary Perspectives of Hostage Reactions, Domestic Abuse and the Stockholm Syndrome," *Australia and New Zealand Journal of Psychiatry* 41, no. 5 (May 2007): 377–84, https://doi.org/10.1080/00048670701261178.

10. Cantor and Price, "Traumatic Entrapment."

11. Helen Reiss, *The Empathy Effect: Seven Neuroscience-Based Keys for Transforming the Way We Live, Love, Work, and Connect across Differences* (Boulder, CO: Sounds True, 2018), 119.

12. Sarah-Jayne Blakemore, *Inventing Ourselves: The Secret Life of the Teenage Brain* (New York: Hachette, 2018); Frances Jensen and Amy Ellis Nutt, *The Teenage Brain: A Neuroscientist's Survival Guide to Raising Adolescents and Young Adults* (Toronto: Harper-Collins, 2015); Daniel Siegel, *Brainstorm: The Power and Purpose of the Teenage Brain* (New York: Penguin, 2013); Laurence Steinberg, *Age of Opportunity: Lessons from the New Science of Adolescence* (New York: Houghton Mifflin Harcourt, 2014); David

Walsh, *Why Do They Act That Way? A Survival Guide to the Adolescent Brain for You and Your Teen* (New York: Simon & Schuster, 2004).

13. Bachand and Djak, "Stockholm Syndrome."

14. CBC News, "Former Colleague Defends Ex-Teacher Accused of Sexual Abuse," *CBC*, September 11, 2006, http://www.cbc.ca/news/canada/british-columbia/former -colleague-defends-ex-teacher-accused-of-sex-abuse-1.576574.

15. Janet Steffenhagen, "*School of Secrets*: Filmmakers' Investigation of the Quest Program at Prince of Wales Ran into Walls of Silence While Probing Why It Was Allowed to Happen," *Vancouver Sun*, October 20, 2007, https://www.pressreader.com/canada /vancouver-sun/20071020/282209416492107.

16. Eunice Lee and Melanie Wood, *School of Secrets*, Bossy Boots Productions with Stranger Productions, 2007, https://strangerproductions.ca/projects/school-of-secrets/.

17. Janet Steffenhagen, "Vancouver School District Obeys Order: Releases More Info about Quest Sex Scandal Updated," *Vancouver Sun*, March 17, 2011, http://vancouver sun.com/news/staff-blogs/vancouver-school-district-obeys-order-releases-more-info -about-quest-sex-scandal-updated.

18. Steffenhagen, "Vancouver School District."

19. Steffenhagen, "Filmmakers' Investigation."

20. Katrina Onstad, "The Learning Curve: Sex with a Teacher. What's Really Going on When Girls Hook Up with Their Teachers," *Elle*, May 2, 2007, http://www.elle.com /life-love/sex-relationships/a13774/sex-with-a-teacher/. All following references to Onstad's article are the same unless otherwise specified.

21. Daniel Siegel, *Mindsight: The New Science of Personal Transformation* (New York: Bantam, 2011), 116. All following references in this paragraph are to this page unless otherwise specified.

22. Angela Duckworth, *Grit: The Power of Passion and Perseverance* (New York: HarperCollins, 2016), 193.

23. Michael Merzenich, *Soft-Wired: How the New Science of Brain Plasticity Can Change Your Life* (San Francisco: Parnassus Publishing, 2013), 163.

# Chapter 8

1. Matthew Lieberman, *Social: Why Our Brains Are Wired to Connect* (New York: Random House, 2013), 5.

2. Amy Saltzman, *A Still Quiet Place for Athletes: Mindfulness Skills for Achieving Peak Performance and Finding Flow in Sports and Life* (Oakland, CA: New Harbinger, 2018), 9.

3. Daniel Siegel, *Brainstorm: The Power and Purpose of the Teenage Brain* (New York: Penguin, 2013), 115.

4. Siegel, *Brainstorm*, 114.

5. Britta Hölzel, James Carmody, Mark Vangel, Christina Congleton, Sita Yerramsetti, Tim Gard, and Sarah Lazar, "Mindfulness Practice Leads to Increases in Regional Brain Gray Matter Density," *Psychiatry Research* 191, no. 1 (January 2011): 36–43, https://www.ncbi.nlm.nih.gov/pmc/articles/PMC3004979/.

6. John Ratey, *Spark: The Revolutionary New Science of Exercise and the Brain* (New York: Little, Brown and Company, 2008), 74.

7. Siegel, *Brainstorm*, 113.

8. Laurence Steinberg, *Age of Opportunity: Lessons from the New Science of Adolescence* (New York: Houghton Mifflin Harcourt, 2014), 158.

9. Helen Reiss, *The Empathy Effect: Seven Neuroscience-Based Keys for Transforming the Way We Live, Love, Work, and Connect across Differences* (Boulder, CO: Sounds True, 2018), 96–97.

10. Stan Rodski, *Neuroscience of Mindfulness: The Astonishing Science Behind How Everyday Hobbies Help You Relax, Work More Efficiently and Lead a Healthier Life* (New York: HarperCollins, 2019), 66.

11. Rodski, *Neuroscience of Mindfulness*, 1–5.

12. Rodski, *Neuroscience of Mindfulness*, 21.

13. Shawn Achor, *The Happiness Advantage: The Seven Principles of Positive Psychology That Fuel Success and Performance at Work* (New York: Random House, 2010), 51.

14. Achor, *Happiness Advantage*, 52.

15. Steinberg, *Age of Opportunity*, 288.

16. Saltzman, *A Still Quiet Place*, 10.

17. Daniel Siegel, *The Mindful Brain: Reflection and Attunement in the Cultivation of Well-Being* (New York: W. W. Norton & Company, 2007).

18. George Mumford, *The Mindful Athlete: Secrets to Pure Performance* (Berkeley, CA: Parallax Press, 2016), 107–8.

19. Reiss, *Empathy Effect*, 55.

20. Mumford, *Mindful Athlete*, 106.

21. Mumford, *Mindful Athlete*, 107–8.

22. Rodski, *Neuroscience of Mindfulness*, 22–23.

23. Phil Jackson and Hugh Delehanty, *Sacred Hoops: Spiritual Lessons of a Hardwood Warrior* (New York: Hyperion, 1995), 12.

24. Lori Desautels and Michael McKnight, *Unwritten: The Story of a Living System. A Pathway to Enlivening and Transforming Education* (Deadwood, OR: Wyatt-MacKenzie Publishing, 2016), 15.

25. Mumford, *Mindful Athlete*, 120.

26. Rodski, *Neuroscience of Mindfulness*, 53.

27. Siegel, *Mindful Brain*, 9.

28. Rodski, *Neuroscience of Mindfulness*, 27–33.

29. Mumford, *Mindful Athlete*, 123.

30. Mumford, *Mindful Athlete*, 127.

31. Mumford, *Mindful Athlete*, 127.

32. Frieda Fanni, "Tom Brady's Secret Weapon: BrainHQ," *DynamicBrain*, https://www.dynamicbrain.ca/posts/78/60/Tom-Brady-s-Secret-Weapon-BrainHQ.html.

33. Jerry Lawton, "Harry Kane Trains his BRAIN to Become England's World Cup Hero," *Daily Star*, June 25, 2018, https://www.dailystar.co.uk/news/latest-news/england-world-cup-hero-harry-16864935.

34. Norman Doidge, *The Brain's Way of Healing: Remarkable Discoveries and Recoveries from the Frontiers of Neuroplasticity* (New York: Penguin, 2016), 10.

35. Doidge, *Brain's Way of Healing*, 15.

36. Doidge, *Brain's Way of Healing*, 22.

37. Doidge, *Brain's Way of Healing*, 15.

38. Daniel Coyle, *The Talent Code: Greatness Isn't Born. It's Grown. Here's How* (New York: Random House, 2009), 214.

39. Richard Harris, "US Military Offers BrainHQ Brain Training to All Personnel," *App Developer Magazine*, January 24, 2018, https://appdevelopermagazine.com/us-military-offers-brainhq-brain-training-to-all-personnel/.

40. Doidge, *Brain's Way of Healing*, 17.

41. Doidge, *Brain's Way of Healing*, 13.

42. Merzenich, e-mail correspondence.

43. Merzenich, e-mail correspondence.

# Chapter 9

1. John Ratey and Richard Manning, *Go Wild: Eat Fat, Run Free, Be Social, and Follow Evolution's Other Rules for Total Health and Well-Being* (New York: Little, Brown Spark, 2014), 110.

2. Bonnie Rochman, "Yay for Recess: Pediatricians Say It Is as Important as Math or Reading," *Time Magazine*, December 31, 2012, https://healthland.time.com/2012/12/31/yay-for-recess-pediatricians-say-its-as-important-as-math-or-reading/.

3. Ratey and Manning, *Go Wild*, 105.

4. Merzenich, e-mail correspondence.

5. John Medina, *Brain Rules: 12 Principles for Surviving and Thriving at Work, Home, and School* (Seattle: Pear Press, 2008), 25.

6. Merzenich, e-mail correspondence.

7. Ratey and Manning, *Go Wild*, 104.

8. Stanley Greenspan with Beryl Benderly, *The Growth of the Mind: And the Endangered Origins of Intelligence* (New York: Perseus Books, 1997), 39.

9. Merzenich, e-mail correspondence.

10. Norman Doidge, *The Brain That Changes Itself: Stories of Personal Triumph from the Frontiers of Brain Science* (New York: Penguin, 2007), 251.

11. Doidge, *Brain That Changes Itself*, 252–53.

12. John Ratey, *Spark: The Revolutionary New Science of Exercise and the Brain* (New York: Little, Brown and Company, 2008), 35.

13. Ratey, *Spark*, 71.

14. Ratey and Manning, *Go Wild*, 103.

15. Ratey, *Spark*, 73.

16. Ratey, *Spark*, 78.

17. Ratey, *Spark*, 103.

18. Ratey, *Spark*, 12.

19. Ratey, *Spark*, 15.

20. Ratey, *Spark*, 8.

21. Laurence Steinberg, *Age of Opportunity: Lessons from the New Science of Adolescence* (New York: Houghton Mifflin Harcourt, 2014), 159.

22. Steinberg, *Age of Opportunity*, 163.

23. Ratey, *Spark*, 10.

24. Ratey, *Spark*, 70.

25. Jim O'Sullivan, "The Wussification of America," *The Atlantic*, December 29, 2010, https://www.theatlantic.com/politics/archive/2010/12/the-wussification-of-america/68652/; Jane McManus, "Wussification Has No Place in Sports," *ESPN*, June 17, 2013, https://www.espn.com/espnw/news-commentary/story/_/id/9395861/espnw-wussification-no-place-sports.

26. Brain Injury Law Center, "Teach Believe Inspire Award—Kimberly Archie," https://www.brain-injury-law-center.com/blog/teach-believe-inspire-kimberly-archie/; Kimberly Archie, Solomon Brannan, Tiffani Bright, Jo Cornell, Debbie Pyka, Mary Seau, Cyndy Feasel, Marcia Jenkins, Leanne Pozzobon, and Darren Hamblin, *Brain Damaged: Two-Minute Warning for Parents* (Westlake Village, CA: USA Sport Safety Publishing, 2019).

27. Michael McCann and Austin Murphy, "New Lawsuit Points Finger at Pop Warner for Mismanagement of Head Injuries," *Sports Illustrated*, September 1, 2016, https://www.si.com/nfl/2016/09/01/pop-warner-youth-football-lawsuit-concussions-cte.

28. Irvin Muchnick, "Newsweek Europe Apologizes—to the Smear Artist!—for Facilitating Exposure of 'Concussion' Movie Partner MomsTeam's Smear of CTE Victim," *Concussion Inc.*, January 26, 2016, https://concussioninc.net/?p=10685.

29. Ratey and Manning, *Go Wild*, 111.

30. Ratey, *Spark*, 12.

31. Ratey, *Spark*, 17.

32. Ratey, *Spark*, 21.

33. Ratey, *Spark*, 32–33.

34. Ratey, *Spark*, 24.

35. Ratey, *Spark*, 29–30.

36. Merzenich, e-mail correspondence.

37. Stan Rodski, *Neuroscience of Mindfulness: The Astonishing Science Behind How Everyday Hobbies Help You Relax, Work More Efficiently and Lead a Healthier Life* (New York: HarperCollins, 2019), 13.

38. Ratey, *Spark*, 63.

39. Rodski, *Neuroscience of Mindfulness*, 9.

40. Ratey, *Spark*, 67.

41. Mihaly Csikszentmihalyi, *Flow: The Psychology of Optimal Performance* (New York: Harper, 1990), 96.

42. Csikszentmihalyi, *Flow*, 198.

43. David Eagleman, *Livewired: The Inside Story of the Ever-Changing Brain* (Toronto: Doubleday, 2020), 12.

44. Eagleman, *Livewired*, 41.

45. Eagleman, *Livewired*, 50.

46. Phil Jackson and Hugh Delehanty, *Sacred Hoops: Spiritual Lessons of a Hardwood Warrior* (New York: Hyperion, 1995), 5–6.

47. Rick Hanson and Richard Mendius, *Buddha's Brain: The Practical Neuroscience of Happiness, Love, and Wisdom* (Oakland, CA: New Harbinger, 2009), 7.

48. Ratey and Manning, *Go Wild*, 100.

49. Michael Merzenich, *Soft-Wired: How the New Science of Brain Plasticity Can Change Your Life* (San Francisco: Parnassus Publishing, 2013), 176.

50. Ratey and Manning, *Go Wild*, 102.

51. Ratey and Manning, *Go Wild*, 123.

52. Ratey and Manning, *Go Wild*, 104.

53. Merzenich, *Soft-Wired*, 158–65.

54. Ratey and Manning, *Go Wild*, 118.

55. Merzenich, e-mail correspondence.

56. Merzenich, e-mail correspondence.

57. Ratey and Manning, *Go Wild*, 118.

58. Ratey and Manning, *Go Wild*, 119.

# Chapter 10

1. Lee-Anne Gray, "The Spectrum of Educational Trauma," *Huffington Post*, November 23, 2015, https://www.huffpost.com/entry/the-spectrum-of-education_b_8619536.

2. Lee-Anne Gray, "When Teachers and Coaches Bully . . ." *Huffington Post*, August 7, 2016, https://www.huffpost.com/entry/when-teachers-and-coaches-bully_b_57a7363 2e4b0ccb023729940.

3. Daniel Pink, *A Whole New Mind: Why Right-Brainers Will Rule the Future* (New York: Penguin, 2005), 113–15.

4. Matthew Lieberman, *Social: Why Our Brains Are Wired to Connect* (New York: Random House, 2013), 155–56.

5. Helen Reiss, *The Empathy Effect: Seven Neuroscience-Based Keys for Transforming the Way We Live, Love, Work, and Connect across Differences* (Boulder, CO: Sounds True, 2018), 55.

6. Daniel Siegel, *Mindsight: The New Science of Personal Transformation* (New York: Bantam, 2011), 118.

7. Lee-Anne Gray, *Self-Compassion for Teens: 129 Activities and Practices to Cultivate Kindness* (Eau Claire, WI: PESI, 2017), 16.

8. Reiss, *Empathy Effect*, 11.

9. Izabela Zych, Maria Ttofi, and David Farrington, "Empathy and Callous-Unemotional Traits in Different Bullying Roles: A Systemic Review and Meta-Analysis," *Trauma, Violence, and Abuse* 20, no. 1 (2019), https://journals.sagepub.com /doi/10.1177/1524838016683456.

10. Reiss, *Empathy Effect*, 10.

11. Daniel Siegel, *Brainstorm: The Power and Purpose of the Teenage Brain* (New York: Penguin, 2013), 86.

12. Sarah-Jayne Blakemore, *Inventing Ourselves: The Secret Life of the Teenage Brain* (New York: Hachette, 2018), 106–7.

13. Reiss, *Empathy Effect*, 80.

14. Reiss, *Empathy Effect*, 17.

15. Reiss, *Empathy Effect*, 29.

16. Reiss, *Empathy Effect*, 30.

17. Reiss, *Empathy Effect*, 18.

18. John Ratey and Richard Manning, *Go Wild: Eat Fat, Run Free, Be Social, and Follow Evolution's Other Rules for Total Health and Well-Being* (New York: Little, Brown Spark, 2014), 35.

19. Merzcnich, e-mail correspondence.

20. Merzenich, e-mail correspondence.

21. Ratey and Manning, *Go Wild*, 161–62.

22. Alex Renton, *Stiff Upper Lip: Secrets, Crimes and the Schooling of a Ruling Class* (London: Weidenfeld & Nicholson, 2017).

23. Ratey and Manning, *Go Wild*, 162.

24. Merzenich, e-mail correspondence.

25. Merzenich, e-mail correspondence.

26. Lieberman, *Social*, 11–12.

27. Brené Brown, *Rising Strong: How the Ability to Reset Transforms the Way We Live, Love, Parent, and Lead* (New York: Random House, 2015), 157.

28. Siegel, *Mindsight*, 124.

29. Jeremy Rifkin, *Empathic Civilization: The Race to Global Consciousness in a World in Crisis* (New York: Penguin, 2010), 119.

30. Rifkin, *Empathic Civilization*, 120.

31. Rifkin, *Empathic Civilization*, 120.

32. Rifkin, *Empathic Civilization*, 121.

33. Lieberman, *Social*.

34. Rifkin, *Empathic Civilization*, 119–20.

35. Mary Gordon, *Roots of Empathy: Changing the World Child by Child* (Toronto: Thomas Allen Publishers, 2005), 78.

36. Reiss, *Empathy Effect*, 10–12.

37. Siegel, *Brainstorm*, 50.

38. Rifkin, *Empathic Civilization*, 604.

39. Reiss, *Empathy Effect*, 30.

40. Rifkin, *Empathic Civilization*, 604.

41. Reiss, *Empathy Effect*, 110.

42. Lori Desautels and Michael McKnight, *Unwritten: The Story of a Living System. A Pathway to Enlivening and Transforming Education* (Deadwood, OR: Wyatt-MacKenzie Publishing, 2016), and *Eyes Are Never Quiet: Listening Beneath the Behaviors of Our Most Troubled Students* (Deadwood, OR: Wyatt-MacKenzie Publishing, 2019).

43. Louis Cozolino, *The Social Neuroscience of Education: Optimizing Attachment and Learning in the Classroom* (New York: W. W. Norton, 2012).

44. John Medina, *Brain Rules: 12 Principles for Surviving and Thriving at Work, Home, and School* (Seattle: Pear Press, 2008), 67–69.

45. Reiss, *Empathy Effect*, 23.

46. Reiss, *Empathy Effect*, 23.

47. Reiss, *Empathy Effect*, 25.

48. Reiss, *Empathy Effect*, 32.

49. Reiss, *Empathy Effect*, 33.

50. Lee-Anne Gray, *Educational Trauma: Examples from Testing to the School-to-Prison Pipeline* (London: Palgrave Macmillan, 2019), 23.

51. Laurence Steinberg, *Age of Opportunity: Lessons from the New Science of Adolescence* (New York: Houghton Mifflin Harcourt, 2014), 156.

52. John Ratey, *Spark: The Revolutionary New Science of Exercise and the Brain* (New York: Little, Brown and Company, 2008), 6.

53. Norman Doidge, *The Brain's Way of Healing: Remarkable Discoveries and Recoveries from the Frontiers of Neuroplasticity* (New York: Penguin, 2016), 364.

54. Michael Merzenich, *Soft-Wired: How the New Science of Brain Plasticity Can Change Your Life* (San Francisco: Parnassus Publishing, 2013), 186–87.

55. Doidge, *Brain's Way of Healing*, 310–11.

56. Reiss, *Empathy Effect*, 54.

57. Reiss, *Empathy Effect*, 55.

58. Reiss, *Empathy Effect*, 54.

# Conclusion

1. Merzenich, e-mail correspondence.

2. Rick Hanson and Richard Mendius, *Buddha's Brain: The Practical Neuroscience of Happiness, Love, and Wisdom* (Oakland, CA: New Harbinger, 2009), 16.

3. Hanson and Mendius, *Buddha's Brain*, 217–18.

4. Hanson and Mendius, *Buddha's Brain*, 16.

5. Hanson and Mendius, *Buddha's Brain*, 218.

# References

Achor, Shawn. *The Happiness Advantage: The Seven Principles of Positive Psychology That Fuel Success and Performance at Work*. New York: Random House, 2010.

Amen, Daniel. *Change Your Brain, Change Your Life: The Breakthrough Program for Conquering Anxiety, Depression, Obsessiveness, Lack of Focus, Anger, and Memory Problems*, revised edition. New York: Penguin, 2015.

Anda, Robert. "Overview of the Adverse Childhood Experiences (ACE) Study." *Multnomah County*. https://multco.us/file/37959/download.

Anda, Robert, Vincent Felitti, James Bremner, John Walker, Charles Whitfield, Bruce Perry, Shanta Dube, and Wayne Giles. "The Enduring Effects of Abuse and Related Adverse Experiences in Childhood." *European Archives of Psychiatry and Clinical Neuroscience* 256, no. 3 (April 2006): 174–86. https://www.researchgate.net/publication/275971785_The_Enduring_Effects_of_Abuse_and_Related_Adverse_Experiences_in_Childhood_A_Convergence_of_Evidence_from_Neurobiology_and__Epidemiology.

Andreatta, Britt. "Potential." *TEDx Talk*, July 22, 2014. https://www.youtube.com/watch?v=yXt_70Ak670&ab_channel=TEDxTalks.

Anthes, Emily. "Inside the Bullied Brain: The Alarming Neuroscience of Taunting." *Boston Globe*, November 28, 2010. http://archive.boston.com/bostonglobe/ideas/articles/2010/11/28/inside_the_bullied_brain/.

Anthony, Andrew. "Alex Renton's Study of the Enduring Culture of Abuse at Britain's Elite Schools Makes for Powerful Reading. Review of *Stiff Upper-Lip: Secrets, Crimes, and the Schooling of a Ruling Class*, by Alex Renton." *Guardian*, April 10, 2017. https://www.theguardian.com/books/2017/apr/10/stiff-upper-lip-secrets-crimes-schooling-of-a-ruling-class-alex-renton-book-review.

Archie, Kimberly, Solomon Brannan, Tiffani Bright, Jo Cornell, Debbie Pyka, Mary Seau, Cyndy Feasel, Marcia Jenkins, Leanne Pozzobon, and Darren Hamblin. *Brain Damaged: Two-Minute Warning for Parents*. Westlake Village, CA: USA Sport Safety Publishing, 2019.

Arden, John. *Rewire Your Brain: Think Your Way to a Better Life*. Hoboken, NJ: John Wiley & Sons, 2010.

Arrowsmith-Young, Barbara. *The Woman Who Changed Her Brain: Unlocking the Extraordinary Potential of the Human Mind*. New York: Free Press, 2012.

Australian Government. "Rewire the Brain." *Try, Test, and Learn Initiative at the Ministry of Social Services*. https://www.dss.gov.au/rewire-the-brain.

Avison, Don. "Quest Outdoor Education Program Review." *Vancouver School Board*. http://www.vsb.bc.ca/sites/default/files/publications/Severed%20-%20Quest%20Outdoor%20Education%20Program%20Review%20-%20March%2016%2C%202011.PDF.

Axelrod, Paul. "Banning the Strap: The End of Corporal Punishment in Canadian Schools." *EdCan*, January 6, 2011. https://www.edcan.ca/articles/banning-the-strap-the-end-of-corporal-punishment-in-canadian-schools/.

Aynsley-Green, Al. *The British Betrayal of Childhood: Challenging Uncomfortable Truths and Bringing about Change*. London: Routledge, 2019.

Babiak, Paul, and Robert Hare. *Snakes in Suits: When Psychopaths Go to Work*. New York: HarperCollins, 2007.

Bachand, Charles, and Nikki Djak. "Stockholm Syndrome in Athletics: A Paradox." *Children Australia* 43, no. 3 (June 2018): 1–6. https://doi.org/10.1017/cha.2018.31.

Bahena, Sofía, North Cooc, Rachel Currie-Rubin, Paul Kuttner, and Monica Ng, eds. *Disrupting the School-to-Prison Pipeline*. Boston: Harvard Educational Review, 2012.

Bauer, Patricia. "Damien Chazelle: American Director and Screenwriter." *Encyclopedia Britannica*. Last updated January 15, 2021. https://www.britannica.com/biography/Damien-Chazelle.

BBC. "Canada: 751 Unmarked Graves Found At Residential School." *BBC*, June 24, 2021. https://www.bbc.com/news/world-us-canada-57592243.

Blakemore, Sarah-Jayne. *Inventing Ourselves: The Secret Life of the Teenage Brain*. New York: Hachette, 2018.

Blakemore, Sarah-Jayne, and Uta Frith. *The Learning Brain: Lessons for Education*. Malden, MA: Wiley-Blackwell, 2005.

Blaylock, Jeannie. "Public Can Check Boy Scout 'Perversion Files' for Accused Molesters." *First Coast News*, November 12, 2019. https://www.firstcoastnews.com/article/news/investigations/boy-scouts-sexual-abuse-investigation/77-6b587579-410a-4f37-a9a0-d7f3cb0000b6.

Blumen, Lorna. *Bullying Epidemic: Not Just Child's Play*. Toronto: Camberley Press, 2011.

Bountiful. "Did Child Abuse Turn Marc Lépine into a Killer?" *Bountiful Films*, March 11, 2014. https://bountiful.ca/abuse-turn-marc-lepine-mass-murderer/.

Brain Injury Law Center. "Teach Believe Inspire Award—Kimberly Archie." https://www.brain-injury-law-center.com/blog/teach-believe-inspire-kimberly-archie/.

Brown, Brené. *Rising Strong: How the Ability to Reset Transforms the Way We Live, Love, Parent, and Lead*. New York: Random House, 2015.

Burgo, Joseph. "All Bullies Are Narcissists: Stories of Bullying and Hazing in the News Break Down to Narcissism and Insecurity." *The Atlantic*, November 14, 2013. https://www.theatlantic.com/health/archive/2013/11/all-bullies-are-narcissists/281407/.

Burke, Andrew, and Klaus Miczek. "Stress in Adolescence and Drugs of Abuse in Rodent Models: Role of Dopamine, CRF, and HPA Axis." *Psychopharmacology* 231, no. 8 (2014): 1557–80. 10.1007/s00213-013-3369-1.

Burra, Nicolas, Dirk Kerzel, David Munoz Tord, Didier Grandjean, and Leonardo Cerevolo. "Early Spatial Attention Deployment Toward and Away from Aggressive Voice." *Social Cognitive and Affective Neuroscience* 14, no. 1 (January 2019): 73–80. https://doi.org/10.1093/scan/nsy100.

Caine, Renata, and Geoffrey Caine. "Understanding a Brain-Based Approach to Learning and Teaching." *ASCD*, 1990. http://www.ascd.org/ASCD/pdf/journals/ed_lead/el_199010_caine.pdf.

Cantor, Chris, and John Price. "Traumatic Entrapment, Appeasement and Complex Post-Traumatic Stress Disorder: Evolutionary Perspectives of Hostage Reactions, Domestic Abuse and the Stockholm Syndrome." *Australia and New Zealand Journal of Psychiatry* 41, no. 5 (May 2007): 377–84. https://doi.org/10.1080/00048670701261178.

Castelloe, Molly. "How Spanking Harms the Brain: Why Spanking Should be Outlawed." *Psychology Today*, February 12, 2012. https://www.psychologytoday.com/ca/blog/the-me-in-we/201202/how-spanking-harms-the-brain.

CBC News. "Former Colleague Defends Ex-Teacher Accused of Sexual Abuse." *CBC*, September 11, 2006. http://www.cbc.ca/news/canada/british-columbia/former-colleague-defends-ex-teacher-accused-of-sex-abuse-1.576574.

———. "2nd Quest Teacher Admits to Sex with a Student." *CBC*, October 16, 2008. http://www.cbc.ca/news/canada/british-columbia/2nd-quest-teacher-admits-to-sex-with-student-1.751710.

CDC. "Adverse Childhood Experiences (ACEs)." *Centers for Disease Control and Prevention*. https://www.cdc.gov/violenceprevention/aces/index.html.

Chazelle, Damien. "Before Writing and Directing 'Whiplash,' Damien Chazelle Lived It." *Los Angeles Times*, December 18, 2014. https://www.latimes.com/entertainment/envelope/la-et-mn-whiplash-writers-damien-chazelle-20141218-story.html.

———. "Divide and Conquer: Damien Chazelle on Why You Should Make a Short First." *MovieMaker*, October 9, 2015. https://www.moviemaker.com/damien-chazelle-on-why-you-should-make-a-short-first/.

———. "Six Film-making Tips from Damien Chazelle: The 'La La Land' Director on How to Make it in La La Land." *Film School Rejects*, December 7, 2016. https://filmschoolrejects.com/6-filmmaking-tips-from-damien-chazelle-6f05f190f427/.

Christoffel, Daniel, Sam Golden, and Scott Russo. "Structural and Synaptic Plasticity in Stress-Related Disorders." *Nature Reviews Neuroscience* 22, no. 5 (2011): 535–49. 10.1515/RNS.2011.044.

Clark, Nick. "*Whiplash* Movie Hit with Backlash from Disgruntled Jazz Fans." Review of *Whiplash* by Damien Chazelle. *Independent*, January 23, 2015. https://www.independent.co.uk/arts-entertainment/films/news/jazz-thriller-whiplash-hit-backlash-disgruntled-jazz-fans-9999858.html.

Clear, James. *Atomic Habits: An Easy and Proven Way to Build Good Habits and Break Bad Ones.* New York: Penguin, 2018.

Clementson, Laura, and Gillian Findlay. "'It's Overwhelming': Survivors Create Public List of Catholic Clerics Accused of Sexual Abuse." *CBC.* December 5, 2019. https://www.cbc.ca/news/canada/catholic-sexual-abuse-london-diocese-1.5384217.

Conkbayir, Mine. *Early Childhood and Neuroscience: Theory, Research and Implications for Practice.* London: Bloomsbury Academic, 2017.

Cooperson, David. *The Holocaust Lessons on Compassionate Parenting and Child Corporal Punishment.* Self-published via CreateSpace, 2014.

Copeland, William, Dieter Wolke, Adrian Angold, and Jane Costello. "Adult Psychiatric Outcomes of Bullying and Being Bullied by Peers in Childhood and Adolescence." *JAMA Psychiatry* 70, no. 4 (2013): 419–26. doi:10.1001/jamapsychiatry.2013.504.

Corcoran, John. *The Teacher Who Couldn't Read: One Man's Triumph over Illiteracy.* New York: Kaplan Publishing, 2008.

Coyle, Daniel. *The Talent Code: Greatness Isn't Born. It's Grown. Here's How.* New York: Random House, 2009.

Cozolino, Louis. *The Social Neuroscience of Education: Optimizing Attachment and Learning in the Classroom.* New York: W. W. Norton, 2012.

Cribb, Robert. "Out of Control Amateur Coaches Mentally Abuse Players." *Toronto Star,* July 8, 2010. https://www.thestar.com/sports/hockey/2010/07/08/outofcontrol_amateur_coaches_mentally_abuse_players.html.

———. "Teachers' Bullying Scarred Us Say Student Athletes." *Toronto Star,* March 14, 2015. https://www.thestar.com/news/canada/2015/03/14/teachers-bullying-scarred-us-say-student-athletes.html.

Csikszentmihalyi, Mihaly. *Flow: The Psychology of Optimal Performance.* New York: Harper, 1990.

*CTV W5.* "Personal Foul: Sports Dreams Shattered by Aggressive Coaches." *CTV W5,* March 14, 2015. https://www.ctvnews.ca/video?clipId=569994&playlistId=1.2279107&binId=1.811589&playlistPageNum=1&binPageNum=1.

Culbert, Lori, and Janet Steffenhagen. "Three Teachers in Quest Case Have Had Relationships with the Pupils They Taught." *Vancouver Sun,* September 14, 2006. http://www.pressreader.com/canada/vancouver-sun/20060914/281543696406093.

Damasio, Antonio. *Descartes' Error: Emotion, Reason and the Human Brain.* New York: Penguin, 2005.

Delville, Yvon, Richard Melloni, and Craig Ferris. "Behavioral and Neurobiological Consequences of Social Subjugation During Puberty in Golden Hamsters." *The Journal of Neuroscience* 18, no. 7 (1998): 2667–72. https://doi.org/10.1523/JNEUROSCI.18-07-02667.1998.

Desautels, Lori, and Michael McKnight. *Eyes Are Never Quiet: Listening Beneath the Behaviors of Our Most Troubled Students.* Deadwood, OR: Wyatt-MacKenzie Publishing, 2019.

———. *Unwritten: The Story of a Living System. A Pathway to Enlivening and Transforming Education.* Deadwood, OR: Wyatt-MacKenzie Publishing, 2016.

Dinich, Heather. "Power, Control and Legacy: Bob Knight's Last Days at IU." *ESPN,* November 29, 2018. https://www.espn.com/mens-college-basketball/story/_/id/23017830/bob-knight-indiana-hoosiers-firing-lesson-college-coaches.

Dispenza, Joe. *Breaking the Habit of Being Yourself: How to Lose Your Mind and Create a New One.* Carlsbad, CA: Hay House, 2012.

Doidge, Norman. *The Brain That Changes Itself: Stories of Personal Triumph from the Frontiers of Brain Science.* New York: Penguin, 2007.

———. *The Brain's Way of Healing: Remarkable Discoveries and Recoveries from the Frontiers of Neuroplasticity.* New York: Penguin, 2016.

Dowd, A. A. "*Whiplash* Maestro Damien Chazelle on Drumming, Directing, and J. K. Simmons." *AVClub*, October 15, 2014. https://film.avclub.com/whiplash-maestro-damien-chazelle-on-drumming-directing-1798273033.

Duckworth, Angela. *Grit: The Power of Passion and Perseverance.* New York: Harper-Collins, 2016.

Dweck, Carol. *Mindset: The New Psychology of Success.* New York: Ballantine, 2006.

Eagleman, David. *Livewired: The Inside Story of the Ever-Changing Brain.* Toronto: Doubleday, 2020.

Eat Drink Films. "The Language of Drums: Director Damien Chazelle and Metallica's Lars Ulrich discuss *Whiplash*." *Eat Drink Films*, November 13, 2014. https://eatdrinkfilms.com/2014/11/13/the-language-of-drums-director-damien-chazelle-and-metallicas-lars-ulrich-discuss-whiplash/.

Encina, Gregorio. "Milgram's Experiment on Obedience to Authority." *University of California*, November 15, 2004. https://nature.berkeley.edu/ucce50/ag-labor/7article/article35.htm.

ESPN, multiple contributors. "Inside a Toxic Culture at Maryland Football." *ESPN*, August 10, 2018. https://www.espn.com/college-football/story/_/id/24342005/maryland-terrapins-football-culture-toxic-coach-dj-durkin.

Fanni, Frieda. "Tom Brady's Secret Weapon: BrainHQ." *DynamicBrain*. https://www.dynamicbrain.ca/posts/78/60/Tom-Brady-s-Secret-Weapon-BrainHQ.html.

FBI. "Patty Hearst." *History: Famous Cases and Criminals. FBI.gov.* https://www.fbi.gov/history/famous-cases/patty-hearst.

Felitti, Vincent. "Reverse Alchemy in Childhood: Turning Gold into Lead." *Family Violence Prevention Fund* 8, no. 1 (Summer 2001): 1–4. http://akhouse.org/tarr/docs/HCR21_Position-Paper_Reverse-Alchemy-in-Childhood_V-Felitti.pdf.

Fitz-Gerald, Sean. "Ask a Juilliard Professor: How Real Is *Whiplash*?" *Vulture*, October 17, 2014. http://www.vulture.com/2014/10/ask-an-expert-juilliard-professor-whiplash.html.

Fraser, Jennifer. *Be a Good Soldier: Children's Grief in English Modernist Novels.* Toronto: University of Toronto Press, 2011.

———. "Posture of the Abused Child." *Kids in the House.* November 10, 2015. https://www.kidsinthehouse.com/blogs/dr-jennifer-fraser/posture-of-the-abused-child.

———. *Teaching Bullies: Zero Tolerance on the Court or in the Classroom.* Vancouver: Motion Press, 2015.

———. "When Teachers Sexually Abuse Students." *Edvocate.* April 5, 2018. https://www.theedadvocate.org/when-teachers-sexually-abuse-students/.

———. "Why We Must Refuse to Submit to Bullying." *Kids in the House.* January 7, 2016. https://www.kidsinthehouse.com/blogs/dr-jennifer-fraser/why-we-must-refuse-to-submit-to-bullying.

Garbarino, James, Edna Guttman, and Janis Seeley. *The Psychologically Battered Child.* San Francisco: Jossey Bass, 1986.

George, Frank, and Dan Short. "The Cognitive Neuroscience of Narcissism." *Journal of Brain Behavior and Cognitive Sciences* 1, no. 6 (2018).

Gershoff, Elizabeth, Andrew Grogan-Kaylor, Jennifer Lansford, Lei Chang, Arnaldo Zelli, Kirby Deater-Deckard, and Kenneth Dodge. "Parent Discipline Practices in an International Sample: Associations with Child Behaviors and Moderation by Perceived Normativeness." *Child Development* 81, no. 2 (March 2010): 487–502. https://www.ncbi.nlm.nih.gov/pmc/articles/PMC2888480/pdf/nihms-198378.pdf.

Gervis, Misia, and Nicola Dunn. "The Emotional Abuse of Elite Child Athletes by Their Coaches." *Child Abuse Review* 13, no. 3 (June 24, 2004). https://onlinelibrary.wiley.com/doi/abs/10.1002/car.843.

Gettel, Oliver. "*Whiplash* Director Damien Chazelle on His Real-Life Inspiration." *Los Angeles Times*, November 11, 2014. https://www.latimes.com/entertainment/movies/moviesnow/la-et-mn-whiplash-damien-chazelle-real-life-inspiration-20141111-story.html.

Gladwell, Malcolm. "In Plain View." *The New Yorker*, September 24, 2012. http://www.newyorker.com/magazine/2012/09/24/in-plain-view.

Goldberg, Alan. "Coaching Abuse: The Dirty, Not-So-Little Secret in Sports." *Competitive Edge*, April 26, 2015. https://www.competitivedge.com/%E2%80%9Ccoaching-abuse-dirty-not-so-little-secret-sports%E2%80%9D.

Gopnik, Alison. *The Gardener and the Carpenter: What the New Science of Child Development Tells Us about the Relationship between Parents and Children.* New York: Farrar, Straus & Giroux, 2016.

Gordon, Mary. *Roots of Empathy: Changing the World Child by Child.* Toronto: Thomas Allen Publishers, 2005.

Gray, Lee-Anne. *Educational Trauma: Examples from Testing to the School-to-Prison Pipeline.* London: Palgrave Macmillan, 2019.

———. *Self-Compassion for Teens: 129 Activities and Practices to Cultivate Kindness.* Eau Claire, WI: PESI, 2017.

———. "The Spectrum of Educational Trauma." *Huffington Post*, November 23, 2015. https://www.huffpost.com/entry/the-spectrum-of-education_b_8619536.

———. "When Teachers and Coaches Bully . . ." *Huffington Post*, August 7, 2016. https://www.huffpost.com/entry/when-teachers-and-coaches-bully_b_57a73632e4b0ccb023729940.

Greenspan, Stanley, with Beryl Benderly. *The Growth of the Mind: And the Endangered Origins of Intelligence.* New York: Perseus Books, 1997.

Guiora, Amos. *Armies of Enablers: Survivor Stories of Complicity and Betrayal in Sexual Assaults.* Chicago: American Bar Association, 2020.

Hanson, Rick. *Hardwiring Happiness: The New Brain Science of Contentment, Calm, and Confidence.* New York: Penguin, 2013.

Hanson, Rick, and Richard Mendius. *Buddha's Brain: The Practical Neuroscience of Happiness, Love, and Wisdom.* Oakland, CA: New Harbinger, 2009.

Harris, Richard. "US Military Offers BrainHQ Brain Training to All Personnel." *App Developer Magazine*, January 24, 2018. https://appdevelopermagazine.com/us-military-offers-brainhq-brain-training-to-all-personnel/.

Hathaway, Bill. "Past Abuse Leads to Loss of Gray Matter in Brains of Adolescents." *Yale News*, December 5, 2011. https://news.yale.edu/2011/12/05/past-abuse-leads-loss-gray-matter-brains-adolescents-0.

Hölzel, Britta, James Carmody, Mark Vangel, Christina Congleton, Sita Yerramsetti, Tim Gard, and Sarah Lazar. "Mindfulness Practice Leads to Increases in Regional Brain Gray Matter Density." *Psychiatry Research* 191, no. 1 (January 2011): 36–43. https://www.ncbi.nlm.nih.gov/pmc/articles/PMC3004979/.

Jackson, Phil, and Hugh Delehanty. *Sacred Hoops: Spiritual Lessons of a Hardwood Warrior*. New York: Hyperion, 1995.

Jamieson, Amber. "Pace Football Coach Accused of Vicious Abuse by Players." *New York Post*, November 23, 2014. https://nypost.com/2014/11/23/pace-football-coach-abused-players-ex-team-members/.

Jensen, Frances, and Amy Ellis Nutt. *The Teenage Brain: A Neuroscientist's Survival Guide to Raising Adolescents and Young Adults*. Toronto: HarperCollins, 2015.

Jülich, Shirley. "Stockholm Syndrome and Child Sexual Abuse." *Journal of Child Sexual Abuse* 14, no. 3 (2004): 107–29.

Kaye, Don. "Interview: *Whiplash* Director Damien Chazelle." *Den of Geek*, October 9, 2014. https://www.denofgeek.com/movies/interview-whiplash-director-damien-chazelle/.

Kuhn, Thomas. *The Structure of Scientific Revolutions*, third edition. Chicago: University of Chicago Press, 1996.

Lang, Daniel. "The Bank Drama: Four Hostages Were Taken During a Bank Robbery in Stockholm, Sweden in 1973. How Did They Come to Sympathize with Their Captors?" *The New Yorker*, November 25, 1974. https://www.newyorker.com/magazine/1974/11/25/the-bank-drama.

Larøi, Frank, Neil Thomas, André Aleman, Charles Fernyhough, Sam Wilkinson, Felicity Deamer, and Simon McCarthy-Jones. "The Ice in Voices: Understanding Negative Content in Auditory Verbal Hallucinations." *Clinical Psychology Review* 67 (February 2019): 1–10. https://doi.org/10.1016/j.cpr.2018.11.001.

Lawton, Jerry. "Harry Kane Trains his BRAIN to Become England's World Cup Hero." *Daily Star*, June 25, 2018. https://www.dailystar.co.uk/news/latest-news/england-world-cup-hero-harry-16864935.

Lee, Ashley. "'Whiplash': J. K. Simmons, Damien Chazelle, on Whether Torment Leads to Talent." *Billboard*, September 27, 2014. https://www.billboard.com/articles/news/6266541/whiplash-jk-simmons-damien-chazelle-on-whether-torment-leads-to-talent.

Lee, Eunice, and Melanie Wood. *School of Secrets*. Bossy Boots Productions with Stranger Productions, 2007. https://strangerproductions.ca/projects/school-of-secrets/.

Lépine, Monique, and Harold Gagné. *Aftermath*. Toronto: Viking, 2008.

Lewis, Rachel. "What Effect Does Yelling Have on Your Child." *The National*, February 26, 2013. https://www.thenational.ae/lifestyle/family/what-effect-does-yelling-have-on-your-child-1.294037.

Lieberman, Matthew. *Social: Why Our Brains Are Wired to Connect*. New York: Random House, 2013.

Lowder, J. Bryan. "Wailing Against the Pansies: Homophobia in *Whiplash*." *Slate Magazine*, October, 22, 2014. https://slate.com/human-interest/2014/10/why-does -whiplash-damien-chazelles-jazz-movie-contain-so-much-homophobia.html.

Maté, Gabor. *In the Realm of Hungry Ghosts: Close Encounters with Addiction*, revised edition. Toronto: Penguin, 2018.

McCann, Michael, and Austin Murphy. "New Lawsuit Points Finger at Pop Warner for Mismanagement of Head Injuries." *Sports Illustrated*, September 1, 2016. https:// www.si.com/nfl/2016/09/01/pop-warner-youth-football-lawsuit-concussions-cte.

McEvoy, Alan. "Abuse of Power: Most Bullying Prevention Is Aimed at Students. What Happens When Adults Are the Aggressors?" *Teaching Tolerance Magazine* 48 (Fall 2014). https://www.tolerance.org/magazine/fall-2014/abuse-of-power.

McEvoy, Alan, and Molly Smith. "Statistically Speaking: Teacher Bullying Is a Real Phenomenon, but It's Been Hard to Quantify—Until Now." *Teaching Tolerance Magazine* 58 (Spring 2018). https://www.tolerance.org/magazine/spring-2018/statis tically-speaking.

McLeod, Saul. "The Milgram Shock Experiment." *Simply Psychology*. Updated 2017. https://www.simplypsychology.org/milgram.html.

McMahon, Tamsin. "Inside Your Teenager's Scary Brain: New Research Shows Incred- ible Cognitive Potential—and Vulnerability—During Adolescence. For Parents, the Stakes Couldn't be Higher." *Macleans Magazine*, January 4, 2015. http://www .macleans.ca/society/life/inside-your-teenagers-scary-brain/.

McManus, Jane. "Wussification Has No Place in Sports." *ESPN*, June 17, 2013. https:// www.espn.com/espnw/news-commentary/story/_/id/9395861/espnw-wussification -no-place-sports.

McNally, Emily, Paz Luncsford, and Mary Armanios. "Long Telomeres and Cancer Risk: The Price of Cellular Immortality." *The Journal of Clinical Investigation* 129, no. 9 (2019): 3474–81. 10.1172/JCI120851.

Medina, John. *Brain Rules: 12 Principles for Surviving and Thriving at Work, Home, and School*. Seattle: Pear Press, 2008.

Merzenich, Michael. *Soft-Wired: How the New Science of Brain Plasticity Can Change Your Life*. San Francisco: Parnassus Publishing, 2013.

Mickleburgh, Rod. "B.C. Girl Felt 'Flattered' by Teacher's Advances." *Globe and Mail*, October 12, 2006. https://www.theglobeandmail.com/news/national/bc-girl-felt -flattered-by-teachers-advances/article18174531/.

———. "Cult-Like Bonding Sparked Rumours." *Globe and Mail*, October 11, 2006. https://beta.theglobeandmail.com/news/national/cult-like-bonding-sparked-rumours /article1107815/?ref=http://www.theglobeandmail.com&.

———. "Ex-Teacher Ellison Admits to Sex Trysts." *Globe and Mail*, October 25, 2006. https://beta.theglobeandmail.com/news/national/ex-teacher-ellison-admits-sex-trysts /article20415846/?ref=http://www.theglobeandmail.com&.

Miller, Alice. *For Your Own Good: Hidden Cruelty in Child-Rearing and the Roots of Violence*. New York: Farrar, Straus and Giroux, 1983.

Miner, Julianna. "Why 70 Percent of Kids Quit Sports by Age 13." *Washington Post*, June 1, 2016. https://www.washingtonpost.com/news/parenting/wp/2016/06/01 /why-70-percent-of-kids-quit-sports-by-age-13/.

Movie Gal. "Interview with 'Whiplash' Film Maker and Oscar Nominee Damien Chazelle." *TheMovieGal.com*, February 8, 2015. https://www.themoviegal.com /single-post/2015/02/08/interview-with-whiplash-filmmaker-oscar-nominee-damien -chazelle.

Muchnick, Irvin. "Newsweek Europe Apologizes—to the Smear Artist!—for Facilitating Exposure of 'Concussion' Movie Partner MomsTeam's Smear of CTE Victim." *Concussion Inc.*, January 26, 2016. https://concussioninc.net/?p=10685.

Mumford, George. *The Mindful Athlete: Secrets to Pure Performance*. Berkeley, CA: Parallax Press, 2016.

Naumetz, Tim. "One in Five Students Suffered Sexual Abuse at Residential Schools, Figures Indicate." *Globe and Mail*, January 17, 2009. https://beta.theglobeandmail .com/news/national/one-in-five-students-suffered-sexual-abuse-at-residential-schools -figures-indicate/article20440061/?ref=http://www.theglobeandmail.com&.

Nuwer, Rachel. "Coaching Can Make or Break an Olympic Athlete: Competitors at the Most Elite Level Need More than Technical Support." *Scientific American*, August 5, 2015. https://www.scientificamerican.com/article/coaching-can-make-or-break-an -olympic-athlete/.

Olsson, Craig, Rob Mcgee, Sheila Williams, and Shyamala Nada-Raja. "A 32-Year Longitudinal Study of Child and Adolescent Pathways to Well-Being in Adulthood." *Journal of Happiness Studies* 14, no. 3 (June 2013): 1–16. https://www.researchgate .net/publication/257589190_A_32-Year_Longitudinal_Study_of_Child_and _Adolescent_Pathways_to_Well-Being_in_Adulthood.

Onstad, Katrina. "The Learning Curve: Sex with a Teacher. What's Really Going on When Girls Hook Up with Their Teachers." *Elle*, May 2, 2007. http://www.elle.com /life-love/sex-relationships/a13774/sex-with-a-teacher/.

O'Sullivan, Jim. "The Wussification of America." *The Atlantic*, December 29, 2010. https://www.theatlantic.com/politics/archive/2010/12/the-wussification-of -america/68652/.

O'Sullivan, John. "Why Kids Quit Sports." *Changing the Game Project*, May 5, 2015. http://changingthegameproject.com/why-kids-quit-sports/.

Pace, Ian. "Music Teacher Sentenced to 11 Years in Prison as Abuse Film Whiplash Prepares for Oscars." *The Conversation*, February 20, 2015. https://theconversation .com/music-teacher-sentenced-to-11-years-in-prison-as-abuse-film-whiplash-prepares -for-oscars-37786.

Panksepp, Jaak. *Affective Neuroscience: The Foundations of Human and Animal Emotions*. Oxford: Oxford University Press, 1998.

Pelletier, Paul. *The Workplace Bullying Handbook: How to Identify, Prevent and Stop a Workplace Bully*. Vancouver: Diversity Publishing, 2018.

Peritz, Ingrid. "The Awful Echoes of Marc Lépine." *Globe and Mail*, December 6, 2004. https://www.theglobeandmail.com/news/national/the-awful-echoes-of-marc-lepine /article1145087/.

Pink, Daniel. *Drive: The Surprising Truth about What Motivates Us*. New York: Penguin, 2009.

———. *A Whole New Mind: Why Right-Brainers Will Rule the Future*. New York: Penguin, 2005.

Plumb, Jacqui, Kelly Bush, and Sonia Kersevich. "Trauma-Sensitive Schools: An Evidence-Based Approach." *School Social Work Journal* (2016). https://www.semantic scholar.org/paper/Trauma-Sensitive-Schools%3A-An-Evidence-Based-Plumb-Bush/39c27626fdef81b93b57eccfc41309772dbc6f78.

Pruitt, Stanton. "Damien Chazelle's Films and the Consequences of Ambition." *Cultured Vultures*, October 7, 2019. https://culturedvultures.com/damien-chazelles-films-and-the-consequences-of-ambition/.

Quilty-Harper, Conrad. "Damien Chazelle: My Next Film Will Have Less Cymbal Throwing." *GQ*, May 11, 2015. https://www.gq-magazine.co.uk/article/damien-chazelle-interview-whiplash-movie-jazz.

Ratey, John. *A User's Guide to the Brain*. New York: Vintage Books, 2002.

———. *Spark: The Revolutionary New Science of Exercise and the Brain*. New York: Little, Brown and Company, 2008.

Ratey, John, and Richard Manning. *Go Wild: Eat Fat, Run Free, Be Social, and Follow Evolution's Other Rules for Total Health and Well-Being*. New York: Little, Brown Spark, 2014.

Reilly, Steve. "Teachers Who Sexually Abuse Students Still Find Classroom Jobs." *USA Today*, December 22, 2016. https://www.usatoday.com/story/news/2016/12/22/teachers-who-sexually-abuse-students-still-find-classroom-jobs/95346790/.

Reisel, Daniel. "The Neuroscience of Restorative Justice." *TED Talk*, February 2013. https://www.ted.com/talks/dan_reisel_the_neuroscience_of_restorative_justice?language=en.

Reiss, Helen. *The Empathy Effect: Seven Neuroscience-Based Keys for Transforming the Way We Live, Love, Work, and Connect across Differences*. Boulder, CO: Sounds True, 2018.

Renton, Alex. "Abuse in Britain's Boarding Schools: Why I Decided to Confront My Demons." *Guardian*, May 4, 2014. https://www.theguardian.com/society/2014/may/04/abuse-britain-private-schools-personal-memoir.

———. *Stiff Upper Lip: Secrets, Crimes and the Schooling of a Ruling Class*. London: Weidenfeld & Nicholson, 2017.

Rifkin, Jeremy. *Empathic Civilization: The Race to Global Consciousness in a World in Crisis*. New York: Penguin, 2010.

Robinson, Tasha. "Damien Chazelle on What Is and What Isn't Ambiguous in *Whiplash*." *The Dissolve*, October 15, 2014. https://thedissolve.com/features/emerging/787-damien-chazelle-on-what-is-and-isnt-ambiguous-abou/.

Rochman, Bonnie. "Yay for Recess: Pediatricians Say It Is as Important as Math or Reading." *Time Magazine*, December 31, 2012. https://healthland.time.com/2012/12/31/yay-for-recess-pediatricians-say-its-as-important-as-math-or-reading/.

Rodski, Stan. *Neuroscience of Mindfulness: The Astonishing Science Behind How Everyday Hobbies Help You Relax, Work More Efficiently and Lead a Healthier Life*. New York: HarperCollins, 2019.

Romm, Cari. "Rethinking One of Psychology's Most Infamous Experiments." *The Atlantic*, January 28, 2015. https://www.theatlantic.com/health/archive/2015/01 /rethinking-one-of-psychologys-most-infamous-experiments/384913/.

Rubin, Roger, Michael O'Keefe, Christian Red, and Nathaniel Vinton. "Mike Rice's Assistant Coach at Rutgers, Jimmy Martelli Resigns, Following Physical and Verbal Abuse Scandal." *New York Daily News*, April 5, 2013. http://www.nydailynews.com /sports/college/rutgers-assistant-baby-rice-cooked-article-1.1308334.

Saltzman, Amy. *A Still Quiet Place for Athletes: Mindfulness Skills for Achieving Peak Performance and Finding Flow in Sports and Life*. Oakland, CA: New Harbinger, 2018.

Sampson, Todd. "Redesign My Brain." *IMDb*, October 2013. https://www.imdb.com /title/tt3322570/episodes!year-2013&ref_-tt_eps_yr_2013.

Sapakoff, Andrew. "College of Charleston Report Hammers 'Jekyll and Hyde' Verbal Abuse by Coach Doug Wojcik." *Post and Courier*, July 2, 2014. https://www.pos tandcourier.com/sports/college-of-charleston-report-hammers-jekyll-and-hyde-verbal -abuse-by-coach-doug-wojcik/article_6be0a402-b99b-5db3-8d0f-4e2b1ea12e4e. html.

Schmidlin, Charlie. "Interview: Director Damien Chazelle talks 'Whiplash,' Musical Editing and His New 'MGM-style' Musical 'La La Land.'" *IndieWire*, October 10, 2014. https://www.indiewire.com/2014/10/interview-director-damien-chazelle-talks -whiplash-musical-editing-his-mgm-style-musical-la-la-land-271422/.

Scott, A. O. "Drill Sergeant in the Music Room." *New York Times*, October 10, 2014. https://www.nytimes.com/2014/10/10/movies/in-whiplash-a-young-jazz-drummer -vs-his-teacher.html.

Siegel, Daniel. *Brainstorm: The Power and Purpose of the Teenage Brain*. New York: Penguin, 2013.

———. *The Mindful Brain: Reflection and Attunement in the Cultivation of Well-Being*. New York: W. W. Norton & Company, 2007.

———. *Mindsight: The New Science of Personal Transformation*. New York: Bantam, 2011.

Silberman, Steve. *NeuroTribes: The Legacy of Autism and the Future of Neurodiversity*. New York: Penguin, 2015.

Silveira, Sarita., Rutvik Shah, Kate Nooner, Bonnie Nagel, Susan Tapert, Michael Bellis, and Jyoti Mishra. "Impact of Childhood Trauma on Executive Function in Adolescence—Mediating Functional Brain Networks and Prediction of High-Risk Drinking." *Biological Psychiatry*, January 2020.

Sims, David. "The Uncomfortable Message in *Whiplash*'s Dazzling Finale." *The Atlantic*, October 22, 2014. https://www.theatlantic.com/entertainment/archive/2014/10/the -ethics-of-whiplash/381636/.

Smith, Erin, Ali Diab, Bill Wilkerson, Walter Dawson, Kunmi Sobowale, Charles Reynolds, Michael Berk, et al. "A Brain Capital Grand Strategy: Toward Economic Reimagination." *Molecular Psychiatry* 26 (October 2020): 3–22. https://www.nature .com/articles/s41380-020-00918-w.

Starecheski, Laura. "Take the ACE Quiz and Learn What It Does—and Doesn't— Mean." *NPR*, March 2, 2015. https://www.npr.org/sections/health-shots/2015 /03/02/387007941/take-the-ace-quiz-and-learn-what-it-does-and-doesnt-mean.

Steffenhagen, Janet. "*School of Secrets*: Filmmakers' Investigation of the Quest Program at Prince of Wales Ran into Walls of Silence While Probing Why It Was Allowed to Happen." *Vancouver Sun*, October 20, 2007. https://www.pressreader.com/canada /vancouver-sun/20071020/282209416492107.

———. "Vancouver School District Obeys Order: Releases More Info about Quest Sex Scandal Updated." *Vancouver Sun*, March 17, 2011. http://vancouversun.com/news /staff-blogs/vancouver-school-district-obeys-order-releases-more-info-about-quest -sex-scandal-updated.

Steinberg, Laurence. *Age of Opportunity: Lessons from the New Science of Adolescence*. New York: Houghton Mifflin Harcourt, 2014.

Stiernberg, Bonnie. "USA Gymnastics Culture of Abuse Runs Far Deeper Than Larry Nassar." *Inside Hook*, July 20, 2020. https://www.insidehook.com/article/sports/usa -gymnasticss-history-of-abuse.

Summit, Roland. "The Child Sexual Abuse Accommodation Syndrome." *Child Abuse and Neglect* 7 (1983): 177–93. https://www.abusewatch.net/Child%20Sexual%20 Abuse%20Accommodation%20Syndrome.pdf.

Taylor, John. "Behind the Veil: Inside the Mind of Men Who Abuse." *Psychology Today*, February 5, 2013. https://www.psychologytoday.com/us/blog/the-reality -corner/201302/behind-the-veil-inside-the-mind-men-who-abuse.

Teicher, Martin. "Impact of Childhood Maltreatment on Brain Development and the Critical Importance of Distinguishing between the Maltreated and Non-Maltreated Diagnostic Subtypes." *International Society for Neurofeedback and Research* (September 2017). https://drteicher.files.wordpress.com/2017/11/isnr_2017_keynote_teicher .pdf.

———. "Wounds That Won't Heal: The Neurobiology of Child Abuse." *Cerebrum: The Dana Forum on Brain Science* 4, no. 2 (January 2000): 50–67. https://www .researchgate.net/publication/215768752_Wounds_that_time_won't_heal_The _neurobiology_of_child_abuse.

Topel, Fred. "Whiplash: Damien Chazelle on Sadistic Writing." *Mandatory*, October 6, 2014. https://www.mandatory.com/fun/770139-whiplash-damien-chazelle-sadistic -writing.

Vaillancourt, Tracy, Eric Duku, Suzanna Becker, Louise Schmidt, Jeffrey Nicol, Cameron Muir, and Harriet MacMillan. "Peer Victimization, Depressive Symptoms, and High Salivary Cortisol Predict Poor Memory in Children." *Brain and Cognition* 77 (2011): 191–99. https://mimm.mcmaster.ca/publications/pdfs/s2.0 -S0278262611001217-main.pdf.

Van der Kolk, Bessel. *The Body Keeps the Score: Brain, Mind, and Body in the Healing of Trauma*. New York: Penguin, 2015.

Verbeke, Willem, Vim Rietdijk, Wouter van den Berg, Roeland Dietvorst, Loek Worm, and Richard Bagozzi. "The Making of the Machiavellian Brain: A Structural MRI Analysis." *Journal of Neuroscience, Psychology and Economics* 4, no. 4 (2011): 205–16. 10.1037/a0025802.

Walsh, David. *Why Do They Act That Way? A Survival Guide to the Adolescent Brain for You and Your Teen*. New York: Simon & Schuster, 2004.

Ward, Lori, and Jamie Strashin. "Sex Offences against Minors: Investigation Reveals More Than 200 Canadian Coaches Convicted in the Last Twenty Years." *CBC*, February 10, 2019. https://www.cbc.ca/sports/amateur-sports-coaches-sexual-offences-minors-1.5006609.

Westcott, Kathryn. "What Is Stockholm Syndrome?" *BBC News Magazine*, August 21, 2013. https://www.bbc.com/news/magazine-22447726.

Wolff, Alexander. "Is the Era of Abusive College Coaches Finally Coming to an End?" *Sports Illustrated*, September 29, 2015. http://www.si.com/college-basketball/2015/09/29/end-abusive-coaches-college-football-basketball.

———. "Why Does Women's Basketball Have So Many Coaching Abuse Problems?" *Sports Illustrated*, October 1, 2015. https://www.si.com/college/2015/10/01/abusive-coaches-womens-basketball-illinois-matt-bollant.

Wooden, John. "The Difference between Winning and Succeeding." *TED Talk*, February 2001. https://www.ted.com/talks/john_wooden_on_the_difference_between_winning_and_success#t-250931.

Zych, Izabela, Maria Ttofi, and David Farrington. "Empathy and Callous-Unemotional Traits in Different Bullying Roles: A Systemic Review and Meta-Analysis." *Trauma, Violence, and Abuse* 20, no. 1 (2019). https://journals.sagepub.com/doi/10.1177/1524838016683456.

# Index

AAP (American Academy of Pediatrics), 143

abuse: bullying as, xxi–xxii; context of, 22; as cycle, 26–27, 31–33, 84; impact of, 2–8, 11–12, 88–95; as learned behavior, 169–70; in mainstream media, 9–10; normalization of, 29–30; not necessary for greatness, 33; verbal, 22. *See also* adults as abusers; adults as bullies; bullying; learned helplessness; trauma

abusive narcissists, 27

ACEs. *See* Adverse Childhood Experiences study (ACEs)

achievement: abuse not necessary for, 33; empathy as vital tool for, 172–73; fear and, 18, 19; gaslighting and, 30; homophobia and, 22; shame and, 19–20; toughness myths, 19, 50. *See also* stress; *Whiplash* (movie)

Achor, Shawn, 46, 128

action plan for writers of culture, 31–33

adults as abusers: accountability and, 74, 79, 82, 85, 93–94, 114; loyalty-binds and, 81–82, 84–85, 109; reaction to being identified, 120–21. *See also* identification with abuser

adults as bullies: cycle of abused children continued with, 31–33; examples of

educators, xvii–xix; power and, 5–6; teaching children to bully, 60. *See also* obedience to authority

Adverse Childhood Experiences study (ACEs): catalyst for research, 92–93; confirming childhood abuse and midlife disease linkage, 94–95; costs stemming from abused children, 95–96; questions from, 90–92; results of, 88–90. *See also* neuroplasticity

AIC (anterior insular cortex), 25

alchemy, 95

alpha-male baboon study, 165

Amen, Daniel, xxiv

American Academy of Pediatrics (AAP), 143

amygdala, 11–12, 45, 129, 176–77

Anda, Robert, 88–92, 95–96, 144. *See also* Adverse Childhood Experiences study (ACEs)

Andersen, Hans Christian, xxvii–xxviii

Andreatta, Britt, 62–63

Andrew Neiman (character), 18, 20–21, 27–28, 29–30

animal behavioral patterns, 111–12

Ansley-Green, Sir Al, 99

anterior cingulate cortex, 164

anterior insular cortex (AIC), 25

Anthes, Emily, 3–6, 83

anxiety, 5
Archie, Kimberly, 149
Arden, John, 66–69
*Armies of Enablers* (Guiora), 119
Arrowsmith, Eaton, 56–60
Arrowsmith-Young, Barbara, 58–59
arts world, 21–22
associative identity disorder, 26
athletes-for-life goal, 149–51. *See also*
    exercise
atomic habits, 3
attachment fragmentation, 112
axonal myelination, 23
axons, 45

Babiak, Paul, 26
baboon study, 165
Bachand, Charles, 112
"bankrupt notion", 3–4
Bannister, Roger, 51
BDNF (brain-derived neurotropic factor),
    69–70, 147
*Be a Good Soldier* (Fraser), xxvii, 86
Bejerot, Nils, 107
Benson, Herbert, 128–29
Bergman, Waltraut, 107
Blakemore, Sarah-Jayne, xxvi, 45, 57
blame, 39, 79
*The Body Keeps the Score* (van der Kolk),
    80–81
borderline personality disorder: about,
    26–27; of author, 115–16, 121; of
    hostages, 109, 110–11; in *Whiplash,* 31
box breathing, 125
Brady, Tom, 133
brain: assessments of, 102; cortisol and,
    xii, xiii–xiv, 3, 23, 126; damaged from
    abuse, xiv; exercise benefiting, 143–
    47; expansiveness of, 13; importance
    of learning about, 32–33, 56–57; as
    muscle, 15; oxygenation of, 155–57;
    uniqueness of, 13, 45. *See also* cortisol;
    healing; neuroplasticity
brain-derived neurotropic factor (BDNF),
    69–70, 147

brainhq.com, 70, 98, 100, 101, 137
brain structure: amygdala, 11–12, 45,
    129, 176–77; anterior cingulate
    cortex, 164; anterior insular cortex,
    25; cerebral cortex, 45; dendrites,
    23; executive function and, xxiii;
    hippocampus, 126; insula, 164;
    learning about, 44–45; myelin,
    23, 52, 136; periaqueductal gray,
    19–20; prefrontal cortex, 19–20, 25;
    somatosensory cortex, 163–64; ventral
    striatum, 25
brain training: author's experience
    with, 121–22; brainhq.com, 70,
    98, 100, 101, 137; evidence of
    brain's malleability, 8–12;
    Merzenich on, 180–81. *See also*
    mindfulness; neuroplasticity;
    practice and brain changes;
    rewiring of the brain
brick metaphor for moments, 182–83
Broeder, Craig, 150–51
Brown, Brené, xviii, 65, 166
bullying: as child abuse, xxi–xxii;
    experience of author's brother, 1–2,
    56, 58, 181; impact of, 2–8, 11–12,
    24, 88–95; labels given to children,
    59–60; as learned behavior, 169–70;
    outdated paradigm of, 10–11;
    separating from narrative, 65–66;
    terminology, xxiii. *See also* abuse;
    adults as bullies; Fraser-Brown,
    Montgomery; Mind-Bully
bullying paradigm: grieving as act of
    rebellion against, 86; neuroplasticity
    as opposite of, 64; normalization and,
    20–21; as outdated, xv–xvi, 10–11,
    64; Sherman on, 28–30
bully-mentors, 24
Burger, Jerry, 39–40

C2Careers start-up, 159–60, 174
Callegari, Stan, 114–16, 118
cancer metaphor, 76–77, 80, 82
cerebral cortex, 45

change: difficulty of thinking in unfamiliar ways, 48–49; immense discipline needed for, 135–37, 138–39; wiring of brain and, 13–15. *See also* neuroplasticity

Chazelle, Damien: on demanding *versus* demeaning behavior, 20–21; high school experiences, 17–18; survival as most important to, 39. *See also Whiplash* (movie)

childhood: adversity in childhood and addiction, 89; bullying as common in, 95; as sacred time, 131; sadness of, 181–83; speaking up during, 39. *See also* abuse; bullying; obedience to authority

Ciceri, Peter, xix

Clark, Olafsson, 105, 107–9

Clear, James, 3

coaches, 23, 41, 51, 52–53, 72. *See also* Fraser-Brown, Montgomery

"Coaching Can Make or Break an Olympic Athlete" (Nuwer), xxvi

Cognitive Behavioral Therapy, 175

Coyle, Daniel, 51–53

concussions, 6, 148–49

Corcoran, John, 60

corporal punishment, xxi–xxii

cortical real estate, 15–16

cortisol, xii, xiii–xiv, 3, 23, 126

cover-ups of abuse, 164–65

Coyle, Daniel, 132–33

critical thinking, 170–71

Csikszentmihalyi, Mihaly, 153, 154

culture of shaming, 168–69. *See also* shame

cyberbullying, 112

Dallman, Mary, 144

Damasio, Antonio, 94, 143, 144

"Dark Tetrad" of bullying traits, 24–26

dead reckoning, 65–68

dendritic elaboration, 23, 45

depression, 75–76, 96, 141–43. *See also* Ellen's story

Desautels, Lori, 130–31, 171

Descartes' Error, 94, 143, 170

Diamond, Marion, 43

Djak, Nikki, 112

Doidge, Norman: on Arrowsmith approach, 60; on brain neuroplasticity, 8–9, 42, 48, 59; on chronic pain, 134; on healing, xxv; on indoctrination, 41; on language, 176; on nerve branching, 45; on play, 145–46; on use-it-or-lose-it principle, 69

dual brain scanner experiments, 163–64

dual personalities. *See* "Jekyll and Hyde" personalities

Duckworth, Angela, 47, 55, 120

Duncan, Kristy, 118

Dweck, Carol, 47–48, 50, 53, 58

Eagleman, David, 45, 62, 153

Eaton Arrowsmith School, 56–60

*Educational Trauma* (Gray), 160

Ehnmark, Kristin, 105–6, 108–9

Eisenberger, Naomi, xxiii

Ellen's story: adults failing her, 74, 78–80, 85; author's experiences similar to, 113; depression story, 76–77, 80, 82, 96; Facebook posting, 75–76, 97; grief over, 86, 87–88; Mind-Bully and, 80–81, 82–83, 85; pain overtaking, 138; school experiences, 71–75; serotonin dysregulation, 77–78

Ellison, Tom, 113–18

emotional abuse, 91

emotional brains (periaqueductal gray), 19–20

emotional neglect, 91–92

emotional pain, xiii, 124, 136

*The Empathic Civilization* (Rifkin), 169–70

empathic listening: author's training in, 160–62; difficulty of, 174; Empathic-Coach, 52–53, 131, 138, 151; within yourself, 176–77

empathy: benefits of activating, 172–74; children mirroring adults in, 162–63;

deficits in, 25; difficulty of, 174; effects of, 170–74; as innate, 164, 165; as opposite of bullying and abuse, 161–62; power and, 171–72; Roots of Empathy program, 170; shame and, 166–69; varying degrees of, 165–66

*Emperor's New Clothes* (Andersen), xxvii–xxviii

*Edvocate* blog, 118

evolutionary theory research, 111–12

executive function, xxiii

exercise: benefits of, 143–47, 152–54; depression and, 141–43; Naperville (IL) program, 147–48, 149–51; in natural environments *versus* gyms, 155–57; sports *versus* fitness, 147–48

expansiveness of the brain, 13

external rewards, 49

Farias, Joaquin, 175

Felitti, Vincent, 88–92, 94, 144. *See also* Adverse Childhood Experiences study (ACEs)

FGF-2 (growth factor), 146–47

fight-flight-freeze response, 22–23, 127

fixed *vs.* growth mindset, 42, 47–48, 49, 50–51, 58

flow state, 152–53, 154

fMRI (functional magnetic resonance imaging), 9, 11, 124

food consumption, 144

Fraser-Brown, Angus, 55–58, 123–25, 134–37

Fraser-Brown, Montgomery: exercising to heal, 142–43; experiences being bullied, xi–xii, xxv–xxvi, xxvii, 2, 5, 38, 141–42, 154–55; success of, 155

Frith, Uta, xxvi

functional magnetic resonance imaging (fMRI), 9, 11, 124

*The Gardener and the Carpenter* (Gopnik), 58

gaslighting, 30

German shepherd experiments, 55

Gopnik, Alison, 58, 88

Gordon, Mary, 170

Gray, Lee-Anne, 160–61, 176

gray matter, xxii

greatness of individuals. *See* achievement

Greenspan, Stanley, 7, 24, 144

grief, xxvii, 87–88

growing talent step, 51–53

growth factors (FGF-2 and VEGF), 146–47

growth *vs.* fixed mindset, 42, 47–48, 49, 50–51, 58

Guerrero, Alex, 133–34

guilt of victims, 80, 84–85

Guiora, Amos, 119

habits, 46

Haigh, Wendy, 99–103

Hanson, Rick, 55, 68, 103, 151, 181–83

Hare, Robert, 26

Harlow, Harry, 62

The Haven on Gariola Island, 125

healing, xiv–xv, xxv, 51–53, 118–22. *See also* empathic listening; exercise; Mind-Brain-Body; mindfulness; neuroplasticity

health conditions: bullying and abuse correlated to, 88–91, 94–95; childhood trauma and, 92; chronic pain, 123–25, 134–38

Hearst, Patty, 110–11

Hebb, William, 13

hippocampus, 126

Hollander, Matthew, 40

homophobia, 22

hooks, bell, 117

hostages, 81. *See also* identification with abuser; Stockholm Syndrome

Hubel, David, 43

Hull, Dean, 114–16, 118, 175–76

humiliation: of Chazelle, 18, 27, 31; of Fraser-Brown (Montgomery), 72–73; lack of empathy and, 162; normalizing, 38; in Quest program,

103, 112, 114–16, 176; trauma and, 111–12. *See also* shame
hypervigilance of abused individuals, 11–12, 39

identification with abuser: animal studies on, 111–12; child sexual abuse victims and, 111; kidnapping of Patty Hearst, 110–11; Kreditbank story, 105–9, 111; naming the aggressor, 119–20; power and dominance hierarchies, 106, 108, 112–13; in Quest program, 113–18; steps to believe in yourself, 119–22; teaching children about, 118–22. *See also* Stockholm Syndrome
IGF-1, 147
imitators, 30
indoctrination, 41
in-group bias, 172–73
"Inside the Bullied Brain" (Anthes), 2–6
insula, 164
intention in mindfulness, 138
intrinsic rewards, 50

Jackson, Phil, 130–31, 153–54
James, William, 42–43
"Jekyll and Hyde" personalities: about, 23–24, 32–33; as associative identity disorder, 84; childhood abuse and, 26–27; falling for, 72; in Kreditbank story, 106; of Quest program leaders, 113
Julliard School of Music, 28–30
justification, 39

Kandel, Eric, 43
Kane, Harry, 133
Kavli Prize, 44, 97–98
"keepers", 8–9
"Keep on learning" mantra, 47–48
Kempermann, Gerd, 145–46, 147
KFS (Klippel-Feil Syndrome), 63, 124, 135. *See also* Fraser-Brown, Angus
kidnapping of Patty Hearst, 110–11
*Kids in the House* blog, 118

kindness, 166
kinesthetic imagery (visualization), 131
Klippel-Feil Syndrome (KFS), 63, 124, 135. *See also* Fraser-Brown, Angus
Kohut, Heinz, 162–63
Kreditbank story, 105–9, 111
Kuhn, Thomas, 8

Lakota people, 130–31
Lang, Daniel, 107, 110
language, power of, 175–76. *See also* humiliation
Lashley, Karl, 42–43
learned helplessness: Angus Fraser-Brown's story, 55–58, 63–64; German shepherd experiments, 55; neuroplasticity and, 11, 57–61; unlearning and rewiring, 61–62, 65–70
"Learning Curve" (Onstad), 116–18
Lieberman, Matthew, xxiii, 124, 166, 171
listening. *See* empathic listening
Ljungberg, Lennart, 107
Lowder, J. Bryan, 22
loyalty-bind, 81–82, 109. *See also* Stockholm Syndrome
Lundblad, Birgitta, 105–7

Machiavellianism, 24, 25–26
Madaule, Paul, 60–61, 175
magnetic resonance imaging (MRI), 9, 126
manipulative deception. *See* Machiavellianism
Manning, Richard, 156
Martelli, Jimmy, 30
Maté, Gabor, xiii
May, Briana, 135
McEvoy, Alan, xviii, xix
McKnight, Michael, 130–31, 171
McNair, Jordan, 132–33
Medina, John, xii–xiii, 55, 144, 171
Merzenich, Michael: background, xix–xxi; on BDNF, 69–70; on brain's ability to change, xxv, 8, 41, 42–45,

63–64, 96–99; on bullying as mutual tragedy, 24; on depression, 96; on doctors' not checking brain activity, 93–94; on exercise, 145–46; on food consumption, 144; on healing, 175; on innateness of empathy, 164, 165; on myelin, 52; on negative messaging, 121; on practice, 46, 49; programs for brain training, 60, 98–100, 133–34, 136, 180–81; on recovery, 12–16; on serotonin dysregulation, 77–78; on sleepwalking through life, 156; on stressful pregnancies, 88–89; on success, 89–90; on Teicher's study, 7; work as ignored or unknown, 3–4
mice experiments, 145–46, 147
microaggressions, xvi
Milgram, Stanley, 35–41, 51
Miller, Alice, 79
Miller, Anesa, 88
Mind-Brain-Body: aligning, xxv–xxvi, 16, 33; empathy and, 177; exercise and, 144–47, 148–49; shame and, 166–67; voice of knowing and, 157
Mind-Bully: about, 10; attacking internally or externally, 10, 78–79; author's experiences, 103; blaming and, 79; bystander interventions and, 38–40; in Ellen's story, 80–81, 82–83, 85; growing talent to disobey, 51–53; internalizing aggressors as, 38–39, 40; lashing out and destroying self, 31; mindfulness and, 130; ousting by naming, 120; as reactive, 52–53; reflecting on our role in, 33; shame and, 167–68; strengthening healthy networks to combat, 16; in toxic culture, 132–33; unconscious brain states compared to, 126–27
mindfulness: about, 125–27; benefits of, 124–27, 134–38; brain responses to, 128–29; brain training and, 138–39; exercise and, 151–52, 155–57; key aspects of, 129–30; stress and, 151–53
mirroring transference, 162–63

mirror neurons, 41–42, 162
monkeys in empty cages, experimenting with, 62
Moskowitz, Michael, 134–37, 138–39
Mumford, George, 128–33, 134, 177
myelin, 23, 52, 136
myelination of corpus callosum, 7, 51–53

names, recording, 119–20
Naperville (IL) fitness programs, 147–48, 149–51
narcissism, 24–25
Nassar, Larry, 41, 72
natural environments for exercise, 155–57
NCAA coaches, 23
negative labels and neuroplasticity, 62–64
negative messaging, 121
neurodiversity, 57, 61
neurological noise, xxi
Neuromotion sessions, 135
neuroplasticity: brick metaphor, 182–83; change and, 4, 57–61, 179–81; harnessing, 12–16; obedience to authority and, 42–46; as opposite of bullying paradigm, 64; recovery of the brain and, 11, 12–16, 96–100; rehabilitation and, 84; Stronger Brains program, 99–103; tree analogy, 45–46. See also brainhq.com; brain training; empathic listening; exercise; Merzenich, Michael; mindfulness; rewiring of the brain
neuroscience: as antidote to bully epidemic, xxiv–xxv; on bullying and brain injury, xx–xxii; dual brain scanner experiments, 163–64; revealing brain's ability to change, 42–45; study of mirroring transference, 162–63
Neuro Tribes (Silberman), 57
Nicholls, Linda, 125
nimbleness, 156–57
normalization of abuse, 29–30, 33
Nuwer, Rachel, xxvi

obedience to authority: fixed *vs.* growth
mindset and, 42, 47–48, 49, 50–51;
growing talent step, 51–53; Milgram's
study and, 37–42; neuroplasticity
and, 42–46. *See also* Stockholm
Syndrome
Ochberg, Frank, 110
Oldgren, Elisabeth, 105–6, 109
Olsson, Jan-Erik, 105–9
Onstad, Katrina, 116–18
opportunity in mindfulness, 139
outdoor education program (Quest), 103,
112, 113–18, 176
out-groups, 172–73
oxygenation of the brain, 155–57

Pace, Ian, 21, 22
Panksepp, Jaak, 87–88, 96
parasympathetic system, 128–29
passive witnesses, 30
Pavlovian conditioning, 4, 43
Pelletier, Paul, xviii, 61
periaqueductal gray. *See* emotional brains
"The Perils of Obedience" (Milgram), 40
Personalized Empowerment Plan, 102
perspective taking, 171–72
PFC. *See* prefrontal cortexphysical abuse,
91
pineal gland, 45
Pink, Daniel, 160
Posit Science, 99
post-traumatic stress disorder (PTSD),
135–36
power: of adult bullies, 5–6; dominance
hierarchies, 106, 108, 112–13;
empathy and, 171–72; humiliation
and, 115, 176; language and, 175–76;
Personalized Empowerment Plan, 102;
willpower, 31
practice and brain changes, 12–16, 46,
49, 51–53, 69. *See also* rewiring of the
brain
prefrontal cortex (PFC), 19–20, 25, 101–2
prisoners, xxiv. *See also* identification with
abuser; Stockholm Syndrome

psychopathy, 24, 25, 26
PTSD (post-traumatic stress disorder),
135–36
"Pyrrhic" victory, 28

Quest (outdoor education program), 103,
112, 113–18, 175

Raithby, David, 125
Ramón y Cajal, Santiago, 153
Ratey, John, 69–70, 126, 146–48, 156
rebels in Milgram's study, 39
recess, 143
*Redesign My Brain* (film), 15
Reiss, Helen, xix, 10, 49, 95, 112, 161–
63, 170, 176–77
relational aggression, xvi
relentlessness in mindfulness, 138–39
reliability in mindfulness, 139
remodeling of the brain. *See*
neuroplasticity; practice and brain
changes; rewiring of the brain
Renton, Alex, 72, 109, 165
repetition. *See* practice and brain changes;
rewiring of the brain
restoration in mindfulness, 139, 141–43
reverse alchemy, 95–96, 103
rewiring of the brain: BDNF and, 69–70,
147; "Rewire the Brain" program,
103; "use it or lose it" principle, 16,
68–69, 137–38; "what fires together,
wires together", 13–14, 52, 66–68,
91. *See also* neuroplasticity; practice
and brain changes
Rice, Mike, 23, 30
Rifkin, Jeremy, 166–67, 168, 169–70
*Rite of Passage in the Narratives of Dante
and Joyce* (Fraser), xxvi–xxvii
"A Road Beyond Loss" (Miller), 88
Rodski, Stan, xiii, 126, 129–30, 151–52

sadism, 24, 25, 26
Säfström, Sven, 105, 107–9
Salzman, Amy, 125, 128
Sampson, Todd, 15

Sandusky, Jerry, 72
Sapolsky, Robert, 165
*School of Secrets* (documentary), 114
school-to-prison pipeline, xxiv
self-belief, 31
Seligman, Martin, 55, 61, 62
serotonin dysregulation, 77–78
sex abuse, 91, 111
sexual harassment. *See* Ellen's story
shame, 19–20, 166–69, 175–76
Sherman, Michael, 28–30
Sherrington, Charles, 42–43
Siegel, Daniel, 119–20, 128, 131, 151,
    161–62, 166
Silberman, Steve, 57
Sims, David, 28
Singer, Tania, 163–64
single photon emission computed
    tomography (SPECT), 9
SLA (Symbionese Liberation Army),
    110–11
"snakes in suits", 26
social bonding, 166, 171. *See also*
    empathic listening; empathy
social pain, 124, 136
social subjugation in animals, 27
somatosensory cortex, 163–64
*Spark* (Ratey), 69–70, 146
sports *versus* fitness, 147–48
sports world and bullying, 21–22, 23.
    *See also* Fraser-Brown, Montgomery;
    Nassar, Larry
Steinberg, Laurence, xxii, 151
Stockholm Syndrome: attachment
    fragmentation in, 112; author
    suffering from, 116; coining of
    phrase, 107; hostages protecting
    captors, 111; recording names to heal,
    119–20; steps of, 110; will to survive
    and, 111
stress: disease correlated to, 88–91;
    exercise and, 146; impact of, xii–xiv,
    23, 127, 168–69; PTSD, 135–36;
    thoughts about, 151–53
Stronger Brains program, 99–103

*Structure of Scientific Revolutions* (Kuhn),
    8
substance abuse, 61
suicide, 21, 28, 71, 74. *See also* Ellen's
    story
Summit, Roland, 79
Symbionese Liberation Army (SLA),
    110–11
sympathy, 166

"talent hot-beds", 51–53
talent-whisperers, 51, 52–53, 132–33,
    138
*Teaching Bullies* (Fraser), xxvii, 160
Teicher, Martin, 6–7, 26, 83–84
Terry Fletcher (character), 20–21, 24,
    27–30, 31, 84
"Theory of Mind", 171
Tolman, Deborah, 117
Tomatis, Alfred, 176
tone of voice, 176–77
totalitarian regimes, 41
toughness myths, 19, 50
train-the-trainer model, 101
translational neuroplasticity, 44
trauma: corporal punishment and,
    xxii–xxiv; health conditions and, 92;
    humiliation and, 111–12; impact of,
    83–84; prefrontal cortex and, 19–20.
    *See also* abuse; bullying; Ellen's story
tree analogy, 58
"Try, Test & Learn" initiative, 102–3
Twain, Mark, 169

uniqueness of the brain, 13
"use it or lose it" principle, 16, 68–69,
    137–38

Vaillancourt, Tracy, 3, 4
van der Kolk, Bessel, 19–20, 39, 78,
    80–81, 109, 111, 152–53
VEGF (growth factor), 146–47
ventral striatum, 25
victims: blaming, 82, 108; bullies
    claiming to be, 35, 39. *See also*

Chazelle, Damien; Ellen's story; Fraser-Brown, Angus; Fraser-Brown, Montgomery
Virginia Tech study, 150
Virtual Iraq program, 135–36
visual cortical model, 44
visualization (kinesthetic imagery), 131
visual processing disability, 55–60
voice of knowing, 177

walking metaphor, 49
Weinstein, Harvey, 29, 41, 72
welfare dependency, 103
"what fires together, wires together" principle, 13–14, 52, 66–68, 91
*Whiplash* (movie): abuse *versus* leadership training, 28–30; fear from trauma, 18–19; fine line between toughness and abuse, 19; "Jekyll and Hyde" personalities in, 24, 84; significance of title, 28; suicide in, 31; summary, 20–21; as tale of triumph, 27–28, 30–31; written in a fever, 18
Wiesel, Torsten, 43
willpower, 31
Wolpert, Daniel, 155
*The Woman Who Changed Her Brain* (Arrowsmith-Young), 58–59
Wooden, John, 53
working memory, 174
workplace terrorism, xviii
writers of culture, 31–33. *See also* *Whiplash* (movie)

X-ray machines, 9